Teaching the Rhetoric of Resistance

Education, Psychoanalysis, and Social Transformation

Series Editors:
Jan Jagodzinski, University of Alberta
Mark Bracher, Kent State University

The purpose of this series is to develop and disseminate psychoanalytic knowledge that can help educators in their pursuit of three core functions of education:

1) facilitating student learning
2) fostering students' personal development, and
3) promoting prosocial attitudes, habits, and behaviors in students (i.e. those opposed to violence, substance abuse, racism, sexism, homophobia, etc.).

Psychoanalysis can help educators realize these aims of education by providing them with important insights into:

1) the emotional and cognitive capacities that are necessary for students to be able to learn, develop, and engage in prosocial behavior
2) the motivations that drive such learning, development, and behaviors, and
3) the motivations that produce anti-social behaviors as well as resistance to learning and development.

Such understanding can enable educators to develop pedagogical strategies and techniques to help students overcome psychological impediments to learning and development, either by identifying and removing the impediments or by helping students develop the ability to overcome them. Moreover, by offering an understanding of the motivations that cause some of our most severe social problems—including crime, violence, substance abuse, prejudice, and inequality—together with knowledge of how such motivations can be altered, books in this series will contribute to the reduction and prevention of such problems, a task that education is increasingly being called upon to assume.

Radical Pedagogy: Identity, Generativity, and Social Transformation
By Mark Bracher

Teaching the Rhetoric of Resistance: The Popular Holocaust and Social Change in a Post 9/11 World
By Robert Samuels

Teaching the Rhetoric of Resistance

The Popular Holocaust and Social Change in a Post–9/11 World

Robert Samuels

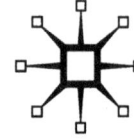

Teaching the Rhetoric of Resistance
Copyright © Robert Samuels, 2007.
All rights reserved. No part of this book may be used or reproduced in any manner whatsoever without written permission except in the case of brief quotations embodied in critical articles or reviews.
First published in 2007 by
PALGRAVE MACMILLAN™
175 Fifth Avenue, New York, NY 10010 and
Houndmills, Basingstoke, Hampshire, England RG21 6XS.
Companies and representatives throughout the world.

PALGRAVE MACMILLAN is the global academic imprint of the Palgrave Macmillan division of St. Martin's Press, LLC and of Palgrave Macmillan Ltd. Macmillan® is a registered trademark in the United States, United Kingdom and other countries. Palgrave is a registered trademark in the European Union and other countries.

ISBN-13: 978-0-2306-0272-4
ISBN-10: 0-2306-0272-X

Library of Congress Cataloging-in-Publication Data

Samuels, Robert.
Teaching the rhetoric of resistance : the popular Holocaust and social change in a post 9/11 world / By Robert Samuels.
   p. cm.
   ISBN 0-230-60272-X (alk. paper)
  1. Holocaust, Jewish (1939–1945)—Study and teaching. 2. Holocaust, Jewish (1939-1945)—Influence. 3. Holocaust, Jewish (1939–1945)—Social aspects. 4. Popular culture—Social aspects. I. Title.
D804.33.S26 2007
940.53'18—dc22   2007018472

A catalogue record of the book is available from the British Library.

Design by Scribe Inc.

First edition: December 2007

10 9 8 7 6 5 4 3 2 1

Transferred to Digital Printing 2011

For Madeleine, with love

Contents

Preface	ix
1 Introduction: Psychoanalytic Pedagogy, Cultural Defense Mechanisms, and Social Change	1
2 What's a Concentration Camp, Dad? Museums, Pedagogy, and the Rhetoric of Popular Culture	29
3 Remembering to Forget: *Schindler's List*, Critical Pedagogy, and the Popular Holocaust	59
4 Life Is Beautiful, but for Whom? Transference, Countertransference, and Student Responses to Teaching about the Holocaust	87
5 Freud Goes to *South Park*: Teaching Against Postmodern Prejudices and Equal Opportunity Hatred	111
6 Teaching Against Binaries: Anti-Semitism and the Holocaust in the Culture of Rhetorical Reversals	131
7 Conclusion: From the Holocaust to the Global War on Terror	149
Notes	155
Works Cited	169
Index	173

PREFACE

My initial focus for this book was a study of how popular culture represents and misrepresents the Holocaust. One of my central concerns was to examine what happens to this historical trauma when it becomes known mainly through movies, television shows, and museums. Moreover, I wanted to study what the renewed interest in the Holocaust during the 1990s said about ethnic identity and trauma during the age of globalization. My original organizing theme in this project was to argue that contemporary audiences share a common set of responses when they encounter the Holocaust through diverse popular media.

In order to further study how audiences respond to the popular representation of the Holocaust and other cultural trauma, I became increasingly interested in examining my own students' reactions to my courses on this historical period. What I then found by concentrating on student responses was that many of their spontaneous reactions fit into four main modes of receiving the messages of the various media. I labeled these four modes idealization, identification, assimilation, and universalization. In turn, I realized that social and psychological forces shaped these reactions and that the field of psychoanalysis offered the best theoretical grounds for detailing these responses. I also discovered that my interest in actual responses from popular audiences was best served by combining ideas from student essays with discussions from popular Web sites, and that by concentrating on what non-academics were saying about the pop culture representations of the Holocaust, I could move away from an overly generalized academic theory of cultural analysis. In other words, by using the real words of my students and discussants on the Web, I could avoid simply projecting academic theories onto the responses of actual individuals.

My interest in engaging students in a discussion of their own responses soon ran into the pedagogical and cultural problem of students resisting critical analysis and self-reflection. Once again, I turned to the field of psychoanalysis to help me understand and work through these resistances, which were often motivated by unconscious and affective aspects. In utilizing the psychoanalytic notion of free association, I developed several pedagogical strategies dedicated to the idea of creating a safe space where students could acknowledge and confront their own unconscious resistances and defense mechanisms. I also realized that the realm of popular culture could be used as a transitional space where students would encounter defense mechanisms in others as a first step in encountering defenses in themselves.

While I was finishing this project, 9/11 happened, and events outside of the classroom soon affected both the contemporary meaning of the Holocaust and my students' resistances to critical analysis. On one level, the Holocaust was being repositioned in the popular culture through the association of the United States and Israel in "The Global War on Terror." This repositioning occurred through the use of political rhetoric, which often posited that America, the world's greatest superpower, had become a victim, and thus all of its acts of retribution were justified. Furthermore, the representation of the United States as an innocent victim was often coupled with a defense of Israel as a historical victim of the Holocaust and a contemporary victim of terrorism. In turn, a new backlash emerged against anyone who was critical of the United States and Israel.

All of these diverse but related issues came to my classroom after a former student at my university decided to start a campaign to pay students to record any faculty member who attacked the United States, Israel, or American foreign policy. This local effort was part of a larger political movement to silence the "tenured radicals" of higher education. Furthermore, conservative activists like David Horowitz and Daniel Pipes attacked any academic critics of the United States and Israel as being, by definition, anti-American and anti-Semitic Holocaust deniers.

In response to these outside political pressures that were making their way into my class discussions, I soon developed a series of pedagogical techniques to allow me to discuss potentially divisive issues in a non-divisive manner. One of my central strategies was to turn to the field of rhetoric in order to define and address the defense mechanisms of my students and the general public. I found that by refusing to name specific political parties, politicians, and political and national affiliations in my class discussions, I was able to prevent students from simply accepting or rejecting the content of my courses solely on the students' preestablished ideological commitments. In short, by using the terms idealization, identification, assimilation, and universalization as rhetorical figures and not as psychological or political processes, I was able to deal with important current issues without naming them and without getting into polarized debates.

This book, then, represents my attempt to combine concerns about history, politics, pedagogy, rhetoric, and popular media while still learning important ethical lessons from the Holocaust. Ultimately, my effort to use a rhetorical and psychoanalytic pedagogy of resistances is centered on my desire to promote critical thinking, engaged citizenship, cultural tolerance, and individual and collective responsibility.

I want to especially thank Mark Bracher for his recent work on pedagogy and psychoanalysis, and Amanda Johnson Moon and Jan Jagodzinski for their support. I would also like to thank my family, friends, colleagues, and students.

Teaching the Rhetoric of Resistance

CHAPTER 1

INTRODUCTION: PSYCHOANALYTIC
PEDAGOGY, CULTURAL DEFENSE
MECHANISMS, AND SOCIAL CHANGE

Traditionally, most efforts to study the Holocaust have been pursued to prevent the reoccurrence of this defining traumatic event in Western culture. The slogan "Never Again" implies that we turn to the Holocaust to learn from it and to prevent its repetition in our own time. Importantly, some of the main lessons educators have sought to derive from the Holocaust are the following:

1. Anti-Semitism and other modes of prejudice can lead to mass murder.
2. People need to fight against intolerance and avoid being passive bystanders.
3. Even democratic people in developed countries can turn to barbaric behavior.
4. The modern Western idea that science and technology always lead to progress is false.
5. Technology can be used to dehumanize and exterminate people.
6. People will act on primal aggressions when they are in a group setting that condones violence.
7. Media technologies can be used to manipulate people through propaganda.
8. The state control of politics, the media, and the military often leads to oppression and death.

9. A higher being will not stop genocide.
10. A humiliated people will act out in violent ways.

All of the lessons listed call for a type of education that is directed toward social change and that critically examines the past in order to analyze and reduce destructive human tendencies in the present and the future. Furthermore, by deriving some good from the tragic loss of human life, these educational efforts are dedicated to honoring those who died in this horrible historical period. Ultimately, the lesson that needs to be learned from the Holocaust is that no one should tolerate the intolerance of others.

Yet, at the start of the twenty-first century, this idea of using education to fight intolerance has itself come under fire. In the contemporary period, the victims of prejudice are often attacked for being victimizers, while the perpetrators of intolerance are at times rhetorically positioned as the true victims.[1] Moreover, educational efforts to teach about racism and other modes of prejudice have come under attack.[2] Thus, in the conservative effort to show how the privileged in our society are the real victims—and the ones who need special protection and such aids as tax cuts—there has been a move to demonize progressive educators.[3] Now, the general culture often celebrates both the tolerance of intolerance and the intolerance of tolerance. In other terms, there has been a major cultural backlash that affects any attempt to use the example of the Holocaust to teach against intolerance and to promote tolerance and other modes of positive social change.

Educators who are still committed to using the study of culture as a way of making society more just are thus forced to situate their efforts in a social and political context that may be hostile to progressive education. Moreover, I have found that students have incorporated, both consciously and unconsciously, many of the antiprogressive messages that circulate in our political and popular cultures. It is therefore necessary for faculty who desire to promote positive social change to start off by acknowledging where students are in terms of their attitudes concerning education, politics, and culture. In fact, a major emphasis of this book is to examine ways in which students and other popular audiences respond to the depiction of the Holocaust and other cultural issues.

While several recent books have discussed the role of the Holocaust in contemporary society, as well as the use of this cultural

INTRODUCTION 3

trauma in diverse educational situations, few of these works have examined how popular audiences respond to the depiction of this historical trauma.[4] Instead, most academic books on this topic have simply ignored popular responses, or they have assumed that all people respond to media and information in the same way as cultural critics.[5] To correct this problem of ignoring popular reactions, I concentrate on discussing diverse comments I have collected in museums, on the Web, and in my university courses. However, I do not simply present these different responses; rather, I have sought to develop an integrated rhetorical theory to help categorize and organize these diverse modes of reception, which often function defensively and on an unconscious level. Moreover, in an effort to interpret and relate these popular responses, I have found it necessary to utilize psychoanalytic theory as a key method for explaining the major defense mechanisms that people employ to respond to trauma and other threatening representations in multiple modes of media.

The main thesis of this work is therefore that an effective way to teach the ethical lessons of the Holocaust in contemporary classrooms is to articulate and overcome the central defense mechanisms shaping students' resistances to critical thinking and positive social change.[6] I also affirm throughout this work that educators need to start off by understanding how students and various popular audiences are actually interpreting the cultural representations of trauma and prejudice. To access what students are thinking and feeling, I endorse an interactive learning environment that values participation and critical analysis. I also claim that this pedagogical strategy is only a first step, however, for I have found that this type of progressive education will fail if teachers do not find a way of engaging and overcoming students' defense mechanisms.

In fact, one reason for teaching about the Holocaust is that this cultural trauma activates people's defense mechanisms in ways that can be used to reveal the ultimate dangers of these defensive reactions.

In reviewing popular responses to the Holocaust in diverse media, I have found that four main types of rhetorical defense mechanisms shape the reception *and* production of trauma in popular culture: idealization, identification, assimilation, and universalization. Furthermore, I show throughout this work that these four responses can be equated in their extreme form with four of the central ethical issues of the Holocaust: the submission to violent authority (idealization); the production of victim identity (identification);

the cultural manipulation of ethnic stereotypes (assimilation); and the passivity of bystanders and collaborators (universalization).

I also posit that these extreme uses of rhetorical defense mechanisms help to shape some of the central forces in contemporary American life: fundamentalism (idealization); identity politics (identification); multiculturalism (assimilation); and globalization (universalization). However, in linking the Holocaust to contemporary society through the shared use of these four cultural defense mechanisms, I do not want to make the false universalizing claim that we are now living in a new period of fascism. Rather, my main objective in presenting this historical connection is to show how educators can employ the critical analysis of the Holocaust to help students move beyond current modes of passivity, prejudice, authoritarianism, and self-denial.

It is essential to stress that I am calling these defenses *rhetorical* in order to emphasize how these psychological mechanisms are mediated by culture and language. Thus, drawing from Aristotle's notion of rhetoric as a public discourse shaped by individual psychology, social hierarchy, and language, I show how the art of persuasion mediates individual psychology and social ideologies.[7] In turn, rhetoric can help teachers move the study of difficult social issues away from purely individual or purely cultural responses: as a mediating force, rhetoric articulates the shared mechanisms of social psychology. Therefore, even though I stress the responses and defense mechanisms of individual students and other popular audiences, I use rhetoric to categorize and understand reoccurring patterns of response.

I also examine these defense mechanisms to expose the ways that contemporary politicians and media employ the rhetoric of trauma to manipulate people. For example, after the attacks of 9/11, the United States' claim of victim status was used to help idealize its own aggressive policies. In this structure, by positioning itself through language and rhetoric as the innocent victim of a traumatic attack (identification), the United States was able to claim moral superiority (idealization) and a self-righteous sense of being beyond criticism. While I do agree that America was the victim of a horrible attack, I think it is important for progressive educators to give their students the rhetorical and psychological tools to question how certain defense mechanisms were used after the assault to justify a whole series of questionable responses. For instance, our society needs to examine how the rhetoric of victimization and universalization were employed to justify a "Global War on Terror" that threatens to be an endless fight with no concrete enemy or battlefield.

In turn, the combination of a universalized war with an identified victim status and an idealized aggressive reaction also triggered a whole series of assimilated cultural stereotypes and prejudices. In other words, our current political world order is being shaped by a group of rhetorical defense mechanisms that need to be addressed by progressive educators. This call for a new model of education is particularly important in light of the fact that this progressive educational effort has itself been successfully demonized by a conservative rhetorical backlash that often circulates in the popular media and shapes many of the unconscious attitudes of our students.

In the face of these political and cultural realities, I argue throughout this book that instead of just ignoring the views, defenses, and desires of our students, educators can develop pedagogical methods that take into account the circulation of prejudices and defense mechanisms in popular culture, political rhetoric, and academic institutions. Nevertheless, I do not think that we can simply address these contemporary issues head on and just try to give our students more accurate information. In fact, psychoanalytic theory tells us that any direct attempt to engage defense mechanisms will only create more resistances. Furthermore, the social factors that shape students' subjectivities are not primarily cognitive; on the contrary, most rhetorical defense mechanisms function on an affective and unconscious level.

Therefore, we need to find indirect ways of addressing students' unconscious emotions, and this is where an integrated theory of psychoanalytic pedagogy and rhetoric comes into play. Since we learn from psychoanalysis that the best way to address resistances is to create a safe space for people to express their past and present unconscious desires and fears, a truly progressive model of pedagogy should develop a nonthreatening learning environment dedicated to free expression and active individual engagement. Yet, we have to keep in mind that educational environments are themselves structured by a whole series of psychodynamic and social influences that block the creation of free discourse. It is therefore necessary for a critical mode of teaching to also develop strategies to transform the traditional roles of students, teachers, and learning while still recognizing the various institutional forces shaping student and teacher subjectivity.

As Mark Bracher argues throughout his *Radical Pedagogy*, a truly transformative educational model would block the idealization of the teacher and the de-idealization of the student (2006, 80–84). Furthermore, Bracher shows why it is necessary to create a space for students to have their unconscious desires and fears recognized even if these attitudes are antithetical to the teacher's values. According to this model of pedagogy, the failure to recognize students' unconscious identities is often caused by the teacher's own self-righteous ideological beliefs concerning proper and improper attitudes. In other words, some teachers would rather stay content feeling that they have the right ideas and students are internalizing the right information than consider how students are actually emotionally reacting to these ideas. Moreover, many progressive teachers want to clearly demonize what they see as negative political and social beliefs in order to show their students what is right and proper. However, when students actually hold the beliefs that the teachers are attacking, the result is usually not a new attitude by the students; instead, students who feel that their identities are being challenged often turn off or actively start to resist the teacher's educational efforts. In turn, when teachers see students resisting, the instructors often blame the students for not caring or trying.

A psychoanalytically informed model of rhetorical pedagogy dedicated to student learning and development thus needs to engage students without shutting them down and attacking their "bad" thoughts and attitudes. One method that has been employed to create a safe space for self-reflection has been the use of nongraded personal writing assignments. In the pedagogy developed by Jeffrey Berman, for instance, students write about a range of sensitive issues without the fear of being judged or graded.[8] While I believe this type of expressivist pedagogy is a step in the right direction, I would argue that most forms of personal writing do not adequately challenge and transform students' rhetorical defense mechanisms. Furthermore, without this type of transformation, I do not think that students can become more active and critical citizens and learners. Therefore, I claim that educators need to find safe and productive ways to get students to acknowledge and work through key defense mechanisms that are blocking learning and critical thinking.

PEDAGOGICAL GOALS

If, as Derek Bok argues in his *Underachieving Colleges*, our shared pedagogical goals are for students to be critical thinkers, engaged learners, ethical human beings, and global citizens, we need a mode of education that is thoughtful, interactive, democratic, and sensitive to cultural differences. However, if one looks at how students are learning in many higher education environments, one often finds passive learners who are sitting in the dark listening to an isolated teacher lecturing on a stage in front of a podium.[9] This type of learning environment may not only train students out of being actively engaged in their educational communities but also function to idealize the expert professor and devalue the ideas of the individual student. In fact, I posit throughout this work that most of our colleges and universities feed the rhetorical defense mechanisms that our students have learned from popular culture, and these defensive structures call for universalized indifference, personal passivity, idealized authority, and assimilated prejudices and stereotypes. In contrast to these rhetorical resistances that are reinforced in higher education, we need to develop learning environments dedicated to overcoming passivity, indifference, authoritarianism, and prejudice.

The main pedagogical tools I endorse in this book are the use of interactive discussions, critical media analysis, nongraded and anonymous writing assignments, in-class free association exercises, computer-mediated group discussions, self-critical essays, and a non-idealizing teacher-student relationship. Furthermore, to work against student indifference and passivity, I affirm that we should develop pedagogical strategies dedicated to helping students confront their shared rhetorical defense mechanisms. The first stage of this process is often the acknowledgment and understanding of rhetorical defenses in others. Thus, I use the critical analysis of popular culture as a safe transitional space where students are taught how to address the reoccurrence of resistances in both the production and the reception of media. However, I do not use popular culture simply to analyze and correct distorted representations of history and society. Instead, I use cultural criticism as a pedagogical tool for critical rhetorical analysis.

One teaching method that I discuss in this work is the strategic employment of nongraded and anonymous writing assignments. These short writing exercises are often centered on having students confront their own rhetorical defense mechanisms and resistances in a non-threatening environment.[10] Another way to help explore these defenses is through the use of anonymous, student-centered online discussions. Like the psychoanalytic use of groups, online discussions can create a space for people to address their own and other people's rhetorical idealizations and identifications. I have also found that discussion forums on the World Wide Web provide a tremendous resource for accessing and examining critically popular defense mechanisms. Once again, the idea here is to use the analysis of other people's defenses and resistances in order to provide a transition to a rhetorical self-analysis. The goal of this process, however, is not to have students confront their own personal traumas and personal histories. The goal is to motivate students to recognize and work through their shared rhetorical resistances to critical analysis and learning.

I argue here that a key factor for allowing students to address these defenses in others and in themselves is the movement away from individual self-reflection toward a more social and rhetorical understanding of resistances. In fact, I posit that the field of rhetoric, especially Aristotle's notion of public rhetoric, helps us see how individual psychology is shaped by larger cultural, social, and linguistic forces. For example, one of the strongest shared defense mechanisms of contemporary politics and culture is the rhetoric of reversed racism. Within this structure, the victims of prejudice are attacked for victimizing the dominant group, while the persecutors are repositioned as being the true victims. We see this rhetorical move when conservatives argue that affirmative action victimizes white males and turns the victims of past prejudices into persecutors. Likewise in the rhetorical backlash against "political correctness," critical intellectuals on the Left are accused of being intolerant fundamentalists because they try to get everyone else to conform to their agenda. In this structure, people who teach tolerance of cultural differences are represented rhetorically as being intolerant, while people are told that they need to be more tolerant of the intolerant rhetoric of others.

A careful analysis of these rhetorical reversals displays how shared defense mechanisms combine psychological, cultural, and linguistic forces, which transcend both purely individual and purely social foundations. Thus, even though I would argue that what drives this conservative rhetorical reversal is the economic desire to justify tax cuts for the wealthy by stigmatizing the welfare state and "liberal" academic discourses, this political movement relies on rhetorical and psychological manipulation. Furthermore, this political rhetorical movement plays on several unconscious psychological mechanisms that are best understood through a psychoanalytic model of subjectivity.

AN INTEGRATED THEORY OF DEFENSE MECHANISMS

In turning to the psychoanalytic theory of defense mechanisms, I examine how these rhetorical devices are considered to be both necessary developmental processes and potentially self-destructive strategies. For instance, in such works as *The Ego and the Id* and *Group Psychology and the Analysis of the Ego*, Sigmund Freud posits that children often react to the traumatic loss of a desired object by replacing the object with an idealized representation of it. On one level, Freud sees this process of idealized identification as a necessary step in children's development of a social conscience, or superego. However, Freud also posits that the failure to let go of a lost object can result in the pathology of melancholia. This conflicting use of idealizing identifications is addressed in Jean Wyatt's *Risking Difference*, where she astutely remarks that Freud's dual interpretation of the same process of identification has led to a whole series of confusions by progressive educators. In fact, Wyatt asserts that while theorists like Judith Butler and Julia Kristeva concentrate on the way the replacement of lost objects with identifications results in a melancholic mode of subjectivity, others like Jacques Lacan and Mikkel Borsh-Jacobson affirm that the self can only develop through identification (11–13). To resolve this ongoing conflict between those who see identification as pathological and those who see it as necessary and developmental, we can follow Wyatt and differentiate among several modes of identification.

Therefore, throughout this work, I concentrate on empathic identifications, idealizing identifications, assimilating identifications, and universalizing identifications. On the most basic level, empathic idealizations use projection to replace the subjectivity of the other with the subjectivity of the self, while idealizing identifications replace the subjectivity of the self with the subjectivity of the other. Likewise, universalizing identifications replace subjectivity with objectivity, while assimilating identifications transform objectivity through cultural relativism. These psychoanalytic processes are thus very similar but also very different in their combinations of subjectivity, objectivity, selfhood, and otherness. I also argue that these identifications help structure the most basic aspects of rhetoric, pedagogy, and contemporary culture. Furthermore, I posit that it is difficult—if not impossible—to work through the fundamental resistances to critical thinking and social change if one does not first differentiate and then integrate these four modes of learning and defending against learning.

Empathic Identifications

Wyatt states that the primary mode of empathic identification can be equated with the basic desire of people to be like other people and to share the same emotions and experiences. However, when this mode of relating becomes an experience of total emotional merging, it can work to deny the separate existence and experiences of others (1). In the case of responding to the Holocaust and other cultural traumas, empathic identification may cause the viewer of a traumatic victimization to feel that he or she has also been traumatized. Here, extreme empathy creates a fixated mode of identity, which in turn effaces the victim's subjectivity. In his *Group Psychology and the Analysis of the Ego*, Freud explains this process by giving the example of a girl in a boarding house who receives a letter from a boy and reacts hysterically, and soon all of the other girls are experiencing the same emotion (49). For Freud, this form of identification shows a hysterical identification with the emotions of others, and we can infer that this symptomatic and empathic merging with the other entails a defense against the loss of a love object. Yet in Freud's example, it is another person's love object that has been lost. In other terms, Freud points to a social and rhetorical mode of desire and empathic identification.

On one level, empathy with another's loss is a necessary social and educational function. Yet in its more extreme or fixated form, empathic identification can work to deny the suffering of others by foregrounding the empathizer's own vicarious experience of victimization. In fact, in contemporary culture we find the development of a mode of cultural identity or identity politics where one bases one's identity on an empathic identification with a founding trauma. For example, many American Jews center their Jewish identity on a remembering of the Holocaust and a sensitivity to anti-Semitism. Once again, this type of empathy can be necessary and effective, but in its more extreme forms, victim identification blocks self-awareness and feeds a masochistic, and at times melancholic, notion of identity. In fact, while I argue throughout this work that it is vital to keep the memory of the Holocaust alive and to guard against anti-Semitic prejudices, I also posit that identity based on trauma and prejudice allows the perpetrators to define the victim's identity and functions to limit the multiple sources for identification. Furthermore, people with victim-based identities may use their real or imaginary suffering to justify their own unjust actions and to shut themselves off from other social problems.

In relation to this fixated form of empathic identification, an important aspect of the psychoanalytic process is to transform people's symptomatic investments in suffering. Therefore, even though one can find a great sense of self-justification and self-identity by seeing oneself as a victim or a member of a victim group, this type of traumatic fixation often represents a blocked attempt at mourning traumatic losses.[11] In fact, in his "Mourning and Melancholia," Freud contrasts the melancholic identification with the lost object to the analytic process of mourning loss and confronting trauma. Following Freud's model, I claim that since new knowledge can challenge the stability of one's prior identity and identifications, a central aspect of critical learning is mourning. In other words, when we ask students to learn something new, we may be introducing them to a traumatic loss of self-identity. In the case of learning about the Holocaust, we are often asking students to give up some of their past idealized understandings of Western culture and human nature itself.[12]

One common reaction to the encounter with lost ideals and lost self-representations is an attempt to defend against this new knowledge by forming an empathic identification with the object of trauma. In this structure, the empathy with the suffering of others may block one's ability to act against suffering in the present and the future: here, the person employing empathic identification resists the types of learning that may promote positive social change. In his several books on the representation and reception of the Holocaust, Dominic LaCapra has returned to this mode of empathy to show

how films like Claude Lanzman's *Shoah* attempt to represent the past by putting the viewer and the maker of the film in the position of "actual reliving, or compulsive acting out of the past—particularly its traumatic suffering—in the present" (*History*, 100). According to LaCapra's insightful criticism, Lanzman appears to be bent on having his witnesses reexperience the past so that Lanzman himself can identify with these traumatized people (101). One reason LaCapra gives for this empathic identification between the person who did witness the traumatic nature of the Holocaust and the filmmaker who did not have the same experiences is that empathic identification gives the artist the ability to become a vicarious victim (111). What LaCapra does not mention here is the psychoanalytic idea that victim identification can provide people with one of the most stable and enduring types of identity.

In his discussion of Lanzman's extreme form of empathic identification, LaCapra also examines Shoshana Felman's pedagogical use of the Holocaust testimonies in her college courses. LaCapra locates in Felman's desire to get her students to experience aspects of the Holocaust a type of acting-out and uncontrolled transference: "Felman's approach to *Shoah* is one of celebratory participation based on empathy or positive transference undisturbed by critical judgment" (111). As is the case with many other contemporary critics that LaCapra discusses, Felman clings to the idea that the Holocaust, like every other trauma, is by definition beyond representation, and so the only thing students can really learn from watching films about this event is the traumatic limits of representation.[13] In relation to Felman's concentration on empathic identification and traumatic fixation, LaCapra locates a tendency of contemporary teachers and theorists to use the Holocaust as a universal symbol of the inability of symbolic representation to represent the real. (111). Furthermore, LaCapra adds that this universalization and absolutization of the traumatic nature of the Holocaust works to block critical analysis and historical distinctions. Thus, instead of examining the particular aspects of the Holocaust, this mode of discourse relies on abstract theoretical claims about the limits of language and symbolic discourse (111). Here, empathic identification is tied to theoretical overgeneralizations with the result of blocking critical understanding.

Another important aspect of LaCapra's critique of Felman's use of empathic identification is his stress on how she tends to promote a mode of self-dramatization that equates the victim of the Holocaust with the witness of trauma and ultimately positions the teacher as a secondary witness. For example, LaCapra notes that in her descriptions of her class on Holocaust testimonies, Felman discusses the ways her students had to live through a "crisis," and thus they themselves became traumatized victims (53). I see this type of vicarious victimhood as one of the major problems with how the Holocaust and other traumatizing subjects are often taught and represented. In addition, this mode of teaching is usually a narcissistic defense mechanism using empathic identification to defend against critical analysis and the understanding of the causes of other people's suffering.

IDEALIZING DEFENSES

In opposition to the empathic identification with traumatic experiences, we find the idealization of authority and aggression. Freud's *Group Psychology and the Analysis of the Ego* presents this type of identification by drawing analogies among the states of being in love, being hypnotized, and being controlled by a political leader (55–58). In each of these instances, Freud finds the same process of reactivating the early relationship between the all-powerful parent and the helpless child. Moreover, Freud insists that all these pathological relationships result in the overestimation of the object of attraction and the replacement of the subject's superego with the superego of the idealized other. Thus, in extreme states of passionate love and political submission, the follower is able to act as if he or she has lost his or her own conscience and has given the role of the superego to the idealized other (56).[14] Furthermore, by tying this idealization of the other to a prior loss of a love object, Freud shows how extreme political authority is often derived from a prior inability to mourn a lost person or ideal.[15]

Freud's theory of idealization can be used to help explain how the failure of Germans to accept their losses in World War I resulted in a collective desire to submit to an idealized authority who could then act as their conscience. Moreover, as post-Freudians like Eric Fromm have posited, people submit to authority figures in order to escape their own responsibility.[16] Idealization can thus be considered to be a central political and subjective defense mechanism, which helps explain not only why people submit to authority but also how people try to escape their own consciences.

In most educational settings, the teacher plays this role of authority figure and the students may seek to rid themselves of any doubt or guilt by projecting ethical and educational responsibility onto the teacher. In turn, many teachers do not challenge this process because they like the feeling of being idealized. We must question, however, how any type of progressive pedagogy can be developed if the students simply idealize the teacher in order to displace their own ethical consciences. It seems that a key to progressive educations should be to fight these defensive processes of identification and idealization, and yet as Bracher stresses in his *Radical Pedagogy*, teachers and students often collude unconsciously to support mutual idealization and thus block critical learning:

> Often, however, the main impediment to education is not the opposition but rather the collusion between the two sets of identity needs, resulting in teachers and students engaging in activities that both parties find supportive of their identities but that do not contribute significantly to realizing the aims of education: learning and identity development. Such pedagogical failures usually go unnoticed, because the primary parties are both satisfied: when students' identities are affirmed, they are happy, produce good teaching evaluations (important forms of recognition for teachers), and remain in school (valuable instances of recognition for administrators)—results that, in turn, produce further recognition for teachers in the form of awards, honors, and salary increases. (79)

Here Bracher shows how idealization may be supported by the reward systems of our educational institutions. Furthermore, we learn from Freud that idealization serves the short-term psychological interests of the teachers and the students, and therefore it is hard to move away from this mutually reinforcing relationship. Yet, it is possible that the only way to teach an ethics of social responsibility is to allow for a de-idealization of authority in order to block the projection of the student's conscience onto the teacher. In fact, I develop throughout this work several pedagogical strategies focused on changing how students and teachers perceive the process and function of idealization.

Following Freud, we need to realize that since empathic and idealized identifications are defense mechanisms aimed to resist the loss of desired objects and identities, any attempt to progress beyond these fixations requires a process of mourning the lost object or ideal. In fact, in his *Changing the Subject in English Class*, Marshall Alcorn posits that every real act of education requires mourning the loss of one's previous ideals and self-representations (6). Thus, we can affirm that progressive educators need to provide a space in which students can work through and remove their investments in previous ideals and desired objects. Importantly, the use of the Holocaust as a subject in this process of cultural mourning can be very effective since this historical trauma is often considered to represent the death of the ideals of modern Western culture. For example, to teach the ethical lessons concerning this event, one has to recognize that the Nazis used modern sciences and technologies to exterminate people, and this knowledge requires a de-idealization of the Western ideal equating scientific discovery with social progress. Moreover, a confrontation with the realities of the Holocaust calls into question many religious notions of an idealized authority figure or caregiver who will intervene to save people from suffering. Yet as Freud posits in *The Future of an Illusion*, the idealization of a deity is one of the hardest infantile wishes to ask people to lose and mourn.

My argument here is not that we should crush all of our students' ideals; rather, we need to help them recognize the different rhetorical defense mechanisms that block the process of learning about new and potentially ego-threatening information. Importantly, one of the only ways students can learn about the foundations of modern culture is if they take a non-idealizing and non-melancholic view of its central attributes. We can understand much about this process of cultural mourning by studying how students respond to the depiction of the Holocaust in popular culture and academic discourse. However, just as Freud posited that resistance is both the key to the blocking of analysis and the only avenue for treatment, we must also affirm that idealization and identification are central defense mechanisms on the royal road of education. Ultimately, a psychoanalytic model of pedagogy affirms that instead of ignoring or hiding the defense mechanisms of students, educational efforts should be centered on exploring and overcoming self-limiting resistances.

Universalization as a Rhetorical Defense Mechanism

To Freud's theories of idealization and identification, I believe it is necessary to add the defensive mechanisms of assimilation and universalization. An understanding of these mechanisms can be derived from Freud's more overtly cultural works. For instance, in his famous descriptions of children's games in *Beyond the Pleasure Principle*, we not only find an account of how children respond to loss and trauma by playing with symbolic representations of lost objects but we also encounter what Freud calls a process of acculturation through repetition. Thus, when Freud watches his grandson throw and then bring back a cotton reel attached to a string, Freud is quick to point out that this child is trying to make up for the absence of his mother by replacing his love object with a symbolic object that can be controlled in an act of symbolic mastery (8–11). Freud also stresses how in this game, the child shouts out the German words for "gone" (fort) and "there" (da) when he throws and retrieves his toy. For Lacan, this game shows how language allows us the illusion of mastering absence through symbolic representation.[17] However, I want to stress that this game can be seen as a defense against loss and trauma and it can be read as a key to the symbolic process of repetition and universalization. Freud himself emphasizes that children like to play the same game and read the same books over and over again, and thus there must be a primary satisfaction in symbolic repetition (10). In fact, it is this enjoyment of symbolic repetition that plays a key role in the rhetoric of universalization. For example, the universalizing ideas that state that we are all victims or that we should all be treated equally under the law are founded on the ability of symbolic representations to create the illusion of repetition and sameness. It is also important to stress that repetition and abstract universality do not exist in nature; rather, repetition and universalization are possible only on the level of rhetoric. Furthermore, we can posit that language gives us the important ability to generalize, yet generalization as a defense mechanism can efface essential differences and create a false sense of sameness.

By stressing the rhetorical and defensive nature of symbolic universalization, I want to emphasize how language helps us master trauma and external reality through symbolic substitution and active repetition. In fact, the major aspects of the modern Enlightenment

INTRODUCTION 17

are based on a universalizing rhetoric of symbolic sameness. For example, the idea that everyone should be treated in the same way under the laws of a democratic society is a modern notion requiring the possibility of universalizing rights and putting every person and group in the same symbolic category. Of course, as Slavoj Zizek has pointed out, an extreme form of this symbolic rhetoric of universality is ethnic cleansing and the denial of all particular cultural and subjective differences in the quest for sameness and identity.

The rhetoric of modern universalization is also a central aspect of the development of Western science, technology, capitalism, and bureaucracy. For, as René Descartes posited, science can exist only if people agree to use the same system of representation (math) and the same method of verification (symbolic repetition).[18] Moreover, Descartes and the other early Enlightenment founders of modern science argued that the first thing one has to do to be scientific is to get rid of all of one's previous beliefs. However, this theory not only pits scientific reason against religious belief but also helps forge the idea of value-free research. That is to say, science creates a social space wherein people can function without values and thus potentially without ethical responsibility. For many contemporary thinkers, this value-free nature of universal science and reason led in its extreme form to the concentration camp.

As Theodor Adorno and Max Horkheimer claim in *Dialectic of Enlightenment*, since science allows for order and control at a distance, it enabled people to participate in the industrial mass extermination of other humans who were treated as pure objects for technological processing. From this perspective, just as the rhetoric of universalization can lead to human rights and human equality, it can also lead to indifference and the creation of a social realm devoid of beliefs and ethical responsibility. Moreover, the extreme mode of universalization in the Holocaust has also been used to explain why many people were able to experience mass extermination as a bureaucratic process based on symbolic control and repetition. Therefore, in the classic example of Karl Adolf Eichmann claiming that he was only doing his bureaucratic job, we see how symbolic work can create a safe space for a responsibility-free zone. It is also important to stress that the universalizing tendencies of the symbolic order of science, bureaucracy, and technology can result in a state of total alienation or self-effacement. By relying on generalized models of symbolic mastery, modern culture develops ways for people to escape unpredictability and the chaos of nature and human interaction. Symbolic repetition also opens the door for people to escape their own subjectivities. For instance, in a footnote to the description of his grandson's game, Freud mentions that the same boy would also stand in front of a mirror and play here and gone by ducking out of sight (11n1). Here, the symbolic replacement of the lost object is generalized to such an extent that the subject him- or herself becomes the lost object.

This loss of the self through universalization can also be equated with the educational stress on modern reason and objectivity. For instance, Parker Palmer posits in his *The Courage to Teach* that the greatest threat to effective pedagogy is the rhetoric of objectivism, which he defines as the idea that truth can be attained only by "disconnecting ourselves, physically and emotionally, from the thing we want to know" (51). What Palmer finds so problematic about most modes of teaching in higher education is that this claim of objectivity is used simply to repress the messy aspects of subjectivity and personal involvement (51). For Palmer the ultimate result of this modern mode of reason is to separate one from the world and to repress one's own subjectivity. What Palmer does not articulate, however, is the psychological and rhetorical defense mechanism that accounts for this type of objectivism.

We can view Freud's grandson's games as specific examples of the general tendency of people to turn to culture, objectivism, and symbolism in order to defend against trauma and loss. For, cultural symbols allow us to universalize negative experiences to such an extent that we become indifferent to our own losses. Furthermore, the social science theory of desensitization through repetition helps us understand why people would choose to reenact painful scenes through symbolic representations. For instance, a central theory of the effect of watching violence in the media is that the sheer repetition of the same symbolic scene works to dissociate the meaning and emotional response of the event from its representation. Here, symbolic representation and universalization result in indifference and self-negation.

I have found that when I try to analyze this role of violence or trauma in the media, the first response of my students is often to say the medium has no meaning, and that one should just sit back and enjoy it. In other words, students defend against analyzing culture by making a universalized statement about popular culture. One way, then, to understand globalized culture is to say that it is a culture of learned indifference: that is, people not only learn how to perceive culture as if it has no value or meaning but also tend to respond to this culture by making universal claims. In the case of studying the popular culture representations of the Holocaust, the denial of meaning and the claims of universal indifference seem to be highly absurd. After all, how can one of the most painful collective human traumas become an object of popular indifference? Moreover, if people are indifferent to the Holocaust, to which types of human suffering are they sensitive? From a Freudian perspective, we can posit that it is not so much that people have become indifferent to human suffering as that symbolic representation distances people from pain and suffering by creating the illusion of symbolic control.

The pop culture representation of the Holocaust is therefore in many ways a contradiction: while a central aspect of our universalizing symbolic culture is the ability to create a sense of remote control through symbolic repetition, the Holocaust was fundamentally a trauma in which control and distance were impossible for the sufferers. This contradiction between the symbolic mode of representation and the traumatic nature of the event itself also structures the central conflict of how to teach about trauma without simply defending against the threatening nature of this event. For instance, one way that some teachers and cultural producers have sought to break away from symbolic distance is to traumatize their audience and try to create an empathic merging between the viewer of trauma and the object of trauma. While I think this process of empathic identification might be necessary to push people to go beyond their own indifference and symbolic distance, I argue in this book that this type of traumatic identification most often blocks the development of critical thinking and educational mourning. Instead of simply trying to break through students' rhetorical defenses, I believe that once students become aware of how they use these defense mechanisms to ward off loss and threatening affects, they can begin to choose to use other cognitive strategies.

The Dialectics of Cultural Assimilation

By seeing defense mechanisms as culturally informed rhetorical strategies, we can gain a better understanding of why students resist learning new information and overcoming old prejudices and stereotypes. In fact, throughout this work I use the concept of assimilation to describe multiple ways in which people in our contemporary culture are motivated to internalize and recirculate oppressive prejudices and stereotypes. One of the central sources for this theory of assimilation comes from Freud's *Jokes and Their Relation to the Unconscious*. On its most basic level, Freud's analysis of humor posits that the first-person joke teller attacks a second-person object in order to bond with a third-party audience. Freud adds that the object of the joke is the enjoyment of the other and that this social mode of bonding works by bribing the audience through enjoyment to tolerate the joke teller's transgressions of social decency (116). In many ways, this theory can serve as a psychoanalytic model for both popular culture and cultural assimilation because we learn from this social psychology that culture allows for the circumvention of censorship and repression through an act of bribing the other with enjoyment.

Jokes can be seen as Freud's way of discussing sublimation as a successful strategy for cultural acceptance; however, as Sander Gilman points out in his *Jewish Self-Hatred*, most of Freud's examples are jokes about Jews that recirculate some of the most vicious anti-Semitic stereotypes. In fact, Gilman posits that Freud's desire to theorize humor is based on his need to assimilate to German culture by substituting a more abstract and universal scientific discourse for his own Jewish language and identity (261). Freud thus turns to Jewish jokes in order to distance himself from his own people and to bond with the dominant group. In other terms, Freud's use of Jewish jokes in his theory mimics his theory of jokes: in both cases, the speaker who comes from a minority group circulates prejudices and stereotypes about his own group in order to bond with the dominant group. Furthermore, Freud adds that one of the fundamental jobs of a joke is to take serious issues and put them in a context where no one takes them seriously: "the joke is merely intended to protect that pleasure from being done away with by criticism" (161). Thus, Freud sees jokes as a way of circumventing social norms by creating a safe outlet for disguised aggression and sexuality, and he also posits that the joke teller bribes the joke's audience with pleasure and that this act of bribing and bonding requires both parties to suspend all criticism (162).

Freud's theory of jokes and his own use of humor can help shed light on why we can consider assimilation as a key rhetorical defense mechanism. Fundamentally, Freud's model of humor is based on the idea that outsiders can be accepted by the dominant social group only if the outsiders make fun of themselves and show that they are willing to sacrifice their own ethnic identities to fit into the dominant group. We therefore need to rethink assimilation as not only the process of generalized social conformity but also the process of representing one's willingness to mock and make fun of one's own identity—while claiming that the mockery has no meaning or value. Assimilation therefore draws from the psychoanalytic defenses of disavowal and negation: in this complicated structure, one repeats prejudices and stereotypes at the same time that one denies their importance and meaning.

In the case of popular culture, assimilation can be seen in the constant recirculation of prejudices as well as in the claim that these prejudices have no value or real meaning. According to Zizek, this type of cynical ideology is based on the structure of a fetish where one both affirms and denies the presence of the same thing (*Sublime*, 28–30). However, I want to show how the fetishistic foundation of assimilation in popular culture is a result of what Freud calls the normal development of a social conscience. For example, in the *Ego and the Id*, Freud constantly returns to the idea that the superego represents a progressive identification with social and educative norms and that this same agency derives its power from being associated with the primal id of instinctual desire (23, 30). In fact, Freud posits that the superego or social conscience is formed when the ego itself is taken as an object that is then judged in comparison with the ideals coming from the dominant social authorities (30). The superego is therefore the agency of assimilation and self-division.

Furthermore, Freud ties the superego to the id to help explain why the ego becomes the slave to this powerful social master. On one level, Freud wants to show that the helplessness of the ego in front of the social ideal is derived from the earliest relationship between instincts and the undeveloped self. On another level, Freud ties the formation of the superego to the primitive identification with the parents during a time when the child is still helpless and dependent (26). In other words, the superego is based on the overlapping of two modes of helplessness: one mode of helplessness comes from our infantile powerlessness in front of internal instincts (the id), another mode comes from the external authority of the parents in relation to the helpless child. Making matters even more complicated is Freud's claim that the superego represents both the presence of primary antisocial instincts and the reaction against these desires (30). Finally, Freud insists that the superego is often unconscious, and consequently people are not even aware of this internal self-division between primary desires and the social reaction against them (54).

The unconscious and divided nature of the superego helps us understand how assimilation functions as a powerful defense mechanism in both popular culture and the reception of media and education. For instance, one of the reoccurring themes of this book is the idea that so many of the people responsible for the circulation of prejudices and stereotypes in popular culture are people who belong to the stigmatized groups. One way of understanding this contemporary phenomenon is through the concept of internalized racism, in which people have accepted stereotypes and prejudices as a source of identity. However, in the case of popular culture, one of the driving forces behind this type of rhetoric is the desire to be popular and thus be accepted by the dominant social order. Therefore, one avenue through which an artist can attain acceptance from a general audience is to represent identities in the most generalized and stereotypical ways. In fact, popular culture tends to rely on a discourse of assimilation and a recirculation of stereotypes and prejudices.

This stress on the "popularity" aspect of popular culture also reveals the connection between assimilation and peer pressure in adolescent culture. In this structure, we find that most students want to be accepted by other students by not only copying the dominant cultural representations of identity but also by internalizing self-hatred. A psychoanalytic theory of assimilation adds to this concept of popularity by stressing the unconscious and divided nature of social subjectivity: people both identify with the dominant group and show an unconscious willingness to sacrifice their own identity for the good of the group they hope to join.

The analysis of popular culture in an educational setting can function to expose this process of assimilation, yet it is also important to stress how assimilation is an unconscious defense mechanism that blocks critical analysis and divides people internally. In fact, I emphasize in this book that most modes of cultural criticism fail to take into account the ways in which the various defense mechanisms like assimilation work to undermine pedagogical efforts. Thus, any progressive educational effort has to provide a method for acknowledging and then working through these resistances. I have found that the first stage in the analysis of assimilation is to recognize on a cognitive level the different components of this social and subjective process. Once one understands how this defense mechanism works in culture and in others, one can begin to examine one's own use of this rhetorical resistance. However, as Freud realized, one cannot simply learn about the unconscious in an educational environment; one needs to learn about one's own unconscious and gain a personal conviction concerning the existence and functioning of shared unconscious mechanisms.

For teachers, this need for students to gain an acceptance of the unconscious produces many stumbling blocks, including the fact that teachers are not trained to analyze their students. While I do not think it is the role of teachers to be psychoanalysts, we cannot ignore the power of the unconscious in the classroom, and so we must find effective ways of engaging the unconscious without turning our classrooms into therapy sessions. One way of approaching this issue is to have students keep a personal journal of their dreams and to have them interpret the symbolism of these unconscious formations on their own. The first stage of this process may be for teachers to use the symbolic analysis of popular culture as a transitional space to model the interpretation of dream elements. Following Freud's method, teachers can show how symbolic elements can be interpreted through free association. Once teachers feel that their students understand the basics of symbolic analysis, it is important to let students experiment with interpreting their own dreams. Some of the key aspects to stress in this rhetorical process are:

1. symbolic elements have both a cultural and a personal association;
2. someone or something in our heads is making creative art films each night;
3. dream elements usually relate to things we don't want to think about;
4. dreams can give us access to repressed truths;
5. desires and fears in dreams are often disguised: nothing is what it appears to be; and
6. everything has to be interpreted.

Most of these basic elements of dream analysis can be modeled through the interpretation of popular culture, but the point is not to stay fixated on the level of culture or individual associations. Instead, this work is dedicated to the idea of making education real and convincing by accessing shared rhetorical mechanisms and desires. As I stress throughout *Teaching the Rhetoric of Resistance*, in order to avoid simply assimilating psychoanalytic theory, cultural criticism, and Holocaust representations as abstract texts, all these areas of discourse need to be tied to a personal and emotional conviction—in short, an identification with the unconscious that is multiple and non-mastered. Students and teachers need to accept the notion that an inner symbolic agency in their subjectivities helps them shape their reactions to others and to their own self-representations. Yet students and teachers should likewise understand that powerful cultural defense mechanisms threaten to turn every important idea and event into an object of non-meaning. To fight this anti-critical process, I posit that educators can first reveal to students the complicated unconscious structures of humor and other modes of postmodern popular culture.

Book Outline

Throughout this work I examine the rhetorical defense mechanisms of idealization, identification, assimilation, and universalization in a wide range of popular cultural products and educational situations. In Chapter 2 I discuss how students and other popular audiences have responded to the depiction of cultural trauma at the United States Holocaust Memorial Museum in Washington, DC. My central goal in this section is to detail the different defense mechanisms and to show how they block a critical understanding of cultural trauma and the promotion of positive social change. I also begin this work by looking at museums in order to emphasize the ways in which popular culture and education are being combined in an unconscious manner in contemporary society. Finally, in positing that museums and other media need to directly confront the cultural prejudices and preconceptions of their viewing audiences, this chapter compares the Holocaust museum in DC to the Museum of Tolerance in Los Angeles. My ultimate objective in this chapter is to display

INTRODUCTION 25

why it is important to develop more interactive, rhetorical, and psychoanalytic models of education and popular culture.

In Chapter 3 I turn my focus to Steven Spielberg's *Schindler's List* in order to provide a pedagogical model for teachers interested in acknowledging and confronting the four central cultural defense mechanisms discussed throughout this book. Instead of simply providing an academic reading of Spielberg's film, I display how this movie can be used as a transitional space for students to first encounter the rhetoric of idealization, identification, assimilation, and universalization in culture and in other people. Since I have found that any direct confrontation with students' own defenses will most often result in a negative reaction, I posit that pop culture can offer the essential initial step in providing a safe space in which students can acknowledge the presence and functions of unconscious resistances.

Chapter 4 utilizes the psychoanalytic notions of idealization, transference, and countertransference as essential pedagogical tools that help students and teachers deal with their own subjective issues concerning the modern concepts of objectivity, science, technology, and education. In using class and Web discussions of Roberto Benigni's *Life Is Beautiful*, I examine how students and teachers can work together to acknowledge and overcome the central resistances to critical analysis and active learning. One of the key aspects of this process is a de-idealizing of the teacher through an exposure of the roles played by the instructor's own defenses in the classroom. This type of self-reflexive teaching is also used to motivate students to engage in a series of self-critical writing exercises.

Chapter 5 situates students' and teachers' defense mechanisms in the contemporary political and cultural context. In examining the film *South Park: Bigger, Longer, and Uncut*, I show how a new mode of postmodern prejudice is shaping the way people inside and outside of educational institutions are dealing with stereotyping, assimilation, and progressive pedagogy. In short, this chapter examines how a rhetorical backlash has successfully reversed postmodern identity politics so that the dominant groups are now considered to be victims, and the minority groups are repositioned to be victimizers. Furthermore, by applying Freud's theory of jokes to a reading of *South Park*, I posit that popular culture motivates students to assimilate regressive political and cultural stereotypes. However, I also argue here that teachers cannot simply attack and condemn these regressive political and cultural moves because students may just reject what they do not agree with and accept what fits into their preconceived worldviews. Therefore, in order to establish a successful progressive pedagogy, I discuss ways of employing a nonbinary and non-self-justifying model of teaching.

Chapter 6 continues my discussion of how to teach about the political manipulation of trauma and prejudice in an age where teachers are often attacked for trying to indoctrinate their students into anti-American, Leftist ideologies. In showing how I examined with my students Peter Novick's *The Holocaust in American Life* and one of David Horowitz's attacks on liberal educators, I discuss a way of using rhetorical analysis to engage in a nonconfrontational method of teaching controversial subjects. This chapter also suggests writing projects that teachers can develop to motivate students to connect the lessons they learn in class to critical self-reflection and the promotion of positive social change.

The conclusion of this work deals with the depiction of trauma and identity in recent events leading from 9/11 to the Global War on Terror. In these current events, I critique the ways the employment of idealization, identification, assimilation, and universalization have been manipulated in order to justify a globalized notion of war and identity politics. Central to my argument is the idea that the popular reaction to the Holocaust serves as an instructive model in helping us understand and then work through the unconscious connections among idealized violence, victim identification, globalized media, and assimilation of internalized prejudices.

Psychoanalytic Pedagogy for Social Change

I do not simply criticize the ways that popular culture depicts the Holocaust and other historical traumas; rather, I emphasize how destructive political and psychological reactions to trauma can be overcome. This work posits that we can help fight prejudice and idealized nationalistic aggression by revealing the processes of denial that allow people to evade the import of the popular culture representations shaping contemporary identifications. That is, we cannot accept the idea that a depiction of the Holocaust is just a movie if people get most of their information about this event from films. Moreover, we need to combat the easy attempts to find in history the ideal character with whom we can idealize in acts of narcissistic identification. This educational task also includes an awareness of the ways in which we project our own rejected aggressive desires and fears onto demonized characters from the past. Furthermore, I posit the importance of being vigilant against using historical trauma and victimization as the foundations for the justification of violent actions.

This educational and cultural process calls for a re-contextualizing of the signs and histories that have been decontextualized by pop culture representations. In order to make students more effective interpreters of popular culture, I posit that we should promote a rhetorical and psychoanalytic mode of media literacy. This type of general education will become increasingly important as we learn more about ourselves and our past through the popular globalized media. Finally, throughout this book I argue that a psychoanalytic cultural studies approach to the analysis of global media helps counter the new forms of prejudice and traumatic denial that threaten all people in contemporary society.

Chapter 2

What's a Concentration Camp, Dad?: Museums, Pedagogy, and the Rhetoric of Popular Culture

In this chapter I explore why educators need to take into account unconscious rhetorical defense mechanisms when they attempt to use the Holocaust and other cultural traumas to teach important lessons concerning cultural tolerance, social responsibility, and personal engagement. By analyzing comments from students and discussants on the Web concerning the United States Holocaust Memorial Museum, I examine the rhetoric of idealization, identification, assimilation, and universalization and how these defense mechanisms function as unconscious resistances to individual and social change. Moreover, in looking at the Los Angeles Museum of Tolerance, I discuss many of the possible ways that educators and other cultural workers can anticipate and work against popular resistances to ethical education. I also use this chapter to outline why the central rhetorical defenses serve to block the importance of the public sphere in contemporary society.

Since this book mainly focuses on how educators can use popular culture to help their students address unconscious rhetorical defense mechanisms that block important ethical and social issues, it is first necessary to clearly articulate the possible relationships between popular culture and progressive pedagogy. In fact, museums offer a vital access to the dialectic between popular culture and modern education since these institutions most often are positioned to be both centers of entertainment and places of cultural education.[1] Furthermore, since 1990 the United States has seen the establishment of many museums dedicated to cultural trauma, and all these institutions highlight their important ethical and educational roles. However, very few people have investigated how museums actually educate and what types of pedagogical strategies they enlist in order to teach the essential ethical lessons they claim to promote.[2] In carefully examining the responses of museum visitors, I show why these institutions are often counterproductive and why a psychoanalytic theory of pedagogy can provide a more effective model for progressive education.

The Public Educational Goals of the United States Holocaust Memorial Museum

According to Edward Linenthal's *Preserving Memory: The Struggle to Create America's Holocaust Museum*, the central designers of the United States Holocaust Memorial Museum in Washington, DC, believed the memorial would: "remind Americans of the dangers of being bystanders, it would teach Americans where Christian anti-Semitism could lead, and it would impress upon Americans the fragile relationship between technology and human values. Some supporters insisted that the museum would provide a crucial lesson in individual responsibility" (66).

From its inception, therefore, the museum sought to promote diverse educational and social goals dedicated to overcoming audience passivity, social prejudices, technological dehumanization, and individual indifference. Linenthal adds that these lessons were coupled with an ambitious public education project:

> Lessons learned would inculcate civic virtue in museum visitors. Ideally, they would emerge from their museum encounter with the Holocaust having a greater appreciation of democracy and a more profound sense of personal commitment to the virtues of pluralism, tolerance, and compromise, and a more sober appreciation of the continuing dangers of anti-Semitism and racism. The implicit message was that the Holocaust clarified the importance of adhering to democratic values,

and offered a historical example of what happened when such values failed. (67)

The goals of the museum are thus the same goals that most progressive educators promoting positive social change endorse, and it will be important to see whether the museum does indeed attain these ends.

It is also essential to look at the various educational strategies the museum designers employed in order to fulfill their educational missions. Before I examine the intentions behind this attempt at civic education, however, I want to start by discussing many of the popular reactions I have collected concerning the Holocaust museum in the nation's capital.

Unlike most other books concerning the reception of the Holocaust in contemporary society, my work concentrates on how popular audiences describe their own encounters with this event. In other terms, I do not commence with an abstract conception of a typical audience response nor do I start off by imposing my own analysis. Instead, I have collected the reactions of nonacademics, and I attempt to place these responses within my general interpretive framework. This ethnographic strategy is based on the idea that many academic books on this topic tend to ignore the question of how people actually respond to popular culture depictions of the Holocaust. Instead, I have found that most academic works on this topic generate a series of ideal audience responses that fit into a particular theoretical format, and this method tends to universalize individual responses and creates a situation where concrete political and educational realities are repressed.[3] By first starting off with individual responses, my work hopes to reverse this tendency and remain close to a pragmatic approach to education and cultural analysis. However, I am aware that this method may alienate some academic readers. In fact, in a review of an earlier version of this book, the reviewer complained that my use of statements from students and people on the Web is unscientific and depends on fringe views instead of the careful analysis of experts. In response to this type of criticism, I want to argue for a mode of analysis that uses the Web and other popular forums as an important resource for collecting the responses and opinions of non-experts.

Rather than seeing the Web as a cultural location where fringe opinions are archived, I believe we can view the Internet as a vital location where we can access popular ideas and ideologies. Moreover, once we collect and organize these responses, we can then compare them to the theories of experts to see whether the projected responses match the actual ways that people are experiencing cultural information. That is to say, instead of simply hypothesizing how audiences internalize cultural representations, I believe we should ask people to respond to or at least study their previous responses. For educators, this ethnographic method entails giving students a voice by not starting off a class with a discussion of the teacher's expert opinion. In fact, psychoanalysis tells us that students will conform to the expectations of the authority figure (the teacher or expert), assimilate the new knowledge to fit into their previous ideas, or simply repress and deny the new information. To help prevent this type of miscommunication, psychoanalysis has developed the idea of free association, which is based on the process of creating a space where people can speak freely without censoring themselves or being censored by others.

On the Web we find many venues where the psychoanalytic concept of free association is approximated. Thus, just as the analyst often helps the process of free association by having the patient face away from the analyst, online forums often create a safe space for free expression by removing the presence of any direct social authority and by allowing the respondent to remain anonymous. However, one should not confuse Internet chat with psychoanalysis because there is no analyst or analytic process present in these virtual forums. Yet, I will turn to the Web throughout this book to show how students and other popular audiences express themselves when they are not being subjected to the direct gaze of a teacher or a social scientist. Therefore, while my study may not be as scientific as some social science surveys, I believe my work gives us access to important information that is often repressed or distorted when it is collected in a more scientific manner. [4]

THE PROBLEMS OF EMPATHIC IDENTIFICATION

One interesting source for ethnographic information is the Web site called "The History Place,"[5] which contains many nonexpert responses to the popular depiction of history. There I found an essay titled "The U.S. Holocaust Museum: Why Christians Should Go" by Barbara Beckwith. This writer begins her article by stressing how she experienced the museum as if she herself were reliving the Holocaust. "The industrial steel elevator I took to start my tour on the fourth floor immediately seemed as confining as the railroad boxcars that carried so many Jews to their death" (1). Like so many other visitors to the memorial, this writer felt a strong empathic identification between herself and the Jewish victims depicted throughout the main exhibit. In fact, her equation of taking a cramped elevator ride with the transportation of the Jews to concentration camps seems to show that the designers of the museum were successful in their desire to have people relate to the Holocaust in a personal and emotional way. However, I believe it is important to question the overall effects of this mode of empathic identification. After all, there is a world of difference between the historical plight of the Jews being sent to their extermination and the visitors of the museum being brought to the next stage of their tour. Furthermore, one of the results of this rhetoric of empathic identification is that the attention of the viewer moves from the suffering of the victims to the emotional experience of the witness, and a possible effect of this transformation may be a privatizing of Holocaust memory. In other words, instead of the museum leading to a greater understanding of the social causes and effects of this cultural trauma, visitors may fixate on their own individual emotional responses.

Like many other popular responses to this museum, Beckwith's response stresses that the main result of her encounter with the depiction of the Holocaust was to make an emotional connection, not to gain more knowledge or cultural understanding. "The Museum did not teach me anything I had not known beforehand. It's not meant to; it's meant to elicit an emotional response, and that it does. It immerses visitors in the experience of the Holocaust" (5). I believe that this response shows some of the power and limits of the use of empathic identification. On the one hand, empathic identification can help people overcome their indifference to an event or person; on the other hand, however, this empathic experience can become centered on the emotional responses of the viewer and not on learning about the lives and possible social lessons of the victims. To be precise, empathic identification cannot be an effective pedagogical tool if it results in a purely personal emotional response. This is not to say that some level of empathic identification is not necessary;

rather, my point is that this type of response should act as a starting point for a more complicated learning experience.

One reason why many people claim that the power of the museum relies on its ability to make an emotional connection with the viewer is that they do not think the museum can actually teach anything new to people who have grown up in a culture full of Holocaust representations. For example, Beckwith states that she did not learn anything at the museum because she already had a great deal of knowledge concerning the Holocaust. "I have read a lot about the Holocaust in history books and novels; I have seen movies like *Exodus* and *Schindler's List*, and television shows and miniseries like *The Holocaust* and *Shoah*. In many ways my imagination is stronger than reality; I had imagined worse" (5). When she writes that her imagination was stronger than reality, and that her knowledge comes from popular culture, we are forced to ask, What reality is she talking about? Does she mean that the simulated depiction of reality that she emotionally reexperienced at the museum was not as strong as her popular culture imagination? Or is she saying that the events of the Holocaust were not really as bad as she imagined them to be? I don't think we can choose between these two alternatives because she herself does not distinguish between the actual events of history and their representation at the museum. After all, how can she make this distinction if the museum has allowed her to relive the actual experience of the Holocaust?

I want to stress here that the problem with this concentration on empathic identification as a way of getting people to connect to the Holocaust is not that it necessarily gets history wrong. Rather, the issue I see is that people like Beckwith are so affected by the simulation of history that they tend to replace the actual event with their own emotional response to the simulation of the event. Thus, by personalizing history through artificial simulation, museums can undermine their one educational mission to teach particular civic and social lessons. Moreover, Beckwith herself reveals one of the central ways that the pedagogical desires of the museum are undermined by its own formal structures: "This is not to say that the museum trivializes the harsh reality of what happened or that it isn't shocking to those unfamiliar with the facts. But the process of making an orderly museum—selecting what will be shown, what will be said, and presenting it tastefully—organizes a reality that at its base defies logic" (5). This insightful comment is structured by the defense mechanism Freud called negation. On one level, Beckwith denies that the museum does indeed trivialize this cultural trauma; on another level, however, she acknowledges the possibility that the narrative and structural organization of the museum may undermine its true meaning and effect. From a psychoanalytic view, this type of contradiction or negation is important because it shows the internalized tension between the recognition and repression of a traumatizing encounter. In this instance, the idea that the museum may actually end up trivializing the Holocaust is so threatening to the viewer that she attempts rhetorically to deny this possibility at the same time she reveals it.

In my many visits to the museum—and in my discussions with students about their encounters with this institution—I often encounter this same type of self-divided rhetoric. For example, visitors often stress how affected they were by the museum, but they also indicate that the organization of the museum made their visit comfortable and manageable. Of course, from a psychoanalytic view, the very definition of trauma is dependent on the idea that people encounter an event that they cannot manage or symbolize, and so one could argue that the way the museum functions to make the encounter comfortable may indicate that the use of media, narrative, and structure at the memorial plays the role of universalizing and desensitizing the viewer. Yet, I would add to this analysis that this process of taming the trauma through distancing and universalizing techniques is coupled with the creation of a substitute affective response through empathic identification. The divided nature of so many popular responses can therefore be explained by this combination of desensitization and resensitization through the rhetorical movement from the encounter with the traumatizing event to the self-reflection on the emotional responses of the viewer.

When I have discussed with students and other academics this idea of self-division in front of the traumatic event, they almost always respond by asking why the depiction of the Holocaust should represent a traumatic encounter for Americans who have no relationship with this historical event. My response to this important question is that the Holocaust inevitably motivates people to consider the extreme inhumanity of humanity, and this encounter with human aggression is presented within the context of our own cultural investments in technology, bureaucracy, propaganda, authority, and prejudice. In fact, I have found that it is virtually impossible for people to discuss the Holocaust without quickly relating this event to their own cultural experiences. Therefore, people not only universalize and identify with this event, but as the following comments will show, they also tend to assimilate their encounter with the depiction of the Holocaust into their own sociocultural contexts.

Assimilating the Trauma of the Other

To help explore this role of assimilation in the popular reactions to the Holocaust museum in Washington, DC, I want to examine an article titled "What They Saw at the Holocaust Museum," in which journalist Philip Gourevitch describes his conversations with students from the Vision Christian Academy of Baltimore after they visited the memorial.[6] One eight- or nine-year-old declared: "The Germans thought they had the right to take over the country because the Jews were different. They were jealous because the Jews were almost ruling the country." With this unintentional repetition of Nazi ideology, we see how people will often absorb new cultural information into their preconceived ideologies. In fact, one of my central points about the role of assimilation as a rhetorical defense mechanisms is that educators—and cultural producers of media with an educational intent—need to take into account the preconceptions of their audience when they attempt to teach important social lessons, and the main way of helping these preconceptions surface is to create a space for interaction and personal feedback.

While it may be hard for museums to be highly interactive, I discuss several ways that the Museum of Tolerance in Los Angeles does try to address the problems of assimilation later in the chapter. For now, though, I want to emphasize the point that an institution like the United States Holocaust Memorial Museum needs to anticipate the ways people will appropriate new cultural information by forcing it into their preexisting social stereotypes. For instance, since many people who come to the museum may know very little about Jewish people or Jewish culture, their first encounter with the multiple representation of "Jews" at the museum could make them feel like the young person discussed earlier who understands why the Germans wanted to get rid of such a powerful group. Of course, the museum cannot anticipate everyone's prejudices and ignorance, but if it were more interactive, it could start a more effective dialogue. Likewise, educators who teach about the Holocaust and other cultural issues can use this example to see why it is so important to create a space where students can safely express their knowledge of and views about a social issue. For example, I often start my courses on the Holocaust with a short, nongraded essay in which students discuss what they already know about the Holocaust. My goal in this process is not to correct them or make them feel ignorant; on the contrary, the idea is to start off by giving students a chance to interact with the subject matter without fear of judgment. I also use their responses to see what level of knowledge students have at the start of the class.

Gourevitch shows that this need to reveal and then interact with people's assimilated views of the Holocaust is vital when he asks students why God let the Holocaust happen to the Jews. One student replied, "They didn't pray," and another student added: "It's a 'jealous God.' Terrible. He's jealous because people worship golden calves, idols." These statements show how previous cultural representations shape how new cultural information is assimilated, and this defensive reaction to a different culture depends on the ideological imposition of social stereotypes. While it may be difficult to shake or change these strong interpretive lenses, educators need to find ways of preventing ideological prejudices from shaping the ways students internalize new information. However, the museum and most academic writers studying the museum do not take into account these preconceptions because they usually work with ideal audiences and not the responses of popular audiences.

My argument here is not simply that we need to pay more attention to how the "average Joe" responds to the depiction of the Holocaust and other cultural trauma. What I am positing is that if educators and institutions really want to promote individual and social change, they need to start off with the actual ideas and feelings of a variety of audiences. In turn, these educators then need to find ways of interacting with these personal responses without activating the various defense mechanisms that block the internalization of new and threatening information. For instance, the need to counter destructive defenses can be seen in the following comment from the teacher of the class that Gourevitch interviewed: "I believe that the Jews are God's chosen people. But they don't recognize that Jesus Christ is the messiah, that He came already. If they had, I think the Lord could have heard their prayers a lot more. In a way, they were praying to a God that they don't really know." The awkward switching between the present and past tenses in this teacher's statement shows how she is assimilating biblical references into her interpretation of the present state of Jewish people. In other terms, the teacher defends against acknowledging the inhumanity of humanity by contextualizing the Holocaust and submitting it to her Christian ideology. While Gourevitch tries to engage this rhetorical defensive reaction, we see that any direct attempt to challenge people's ideologies will most often result in a further strengthening of these defense mechanisms. The psychoanalytic method, then, allows the defenses to emerge and gives people tools to overcome their own defensive strategies.

To see how a direct attempt to question people's defenses and prejudices often works to increase these reactions, we can look at what happens when Gourevitch himself tries to educate this instructor by telling her that her own desire to teach her students about other cultures is undermined by her belief that other religions are misguided. In response to Gourevitch's direct criticism, the teacher states: "It's similar to when we teach about Native Americans. Since we are a Christian school, we recommend that the children pray that the people of that country would come to know Jesus Christ, and that they pray for their needs." A striking aspect of this response is that it assimilates a postmodern rhetorical affirmation of multicultural tolerance into a traditional call for fundamental Christian values. Of course, underlying this ideology is a universalizing claim that all people should recognize Christ as their savior. The prejudices and defensiveness of the teacher reflect on the need for educational institutions, like the Holocaust museum in Washington, DC, to engage students and teachers in a dialogue that corrects misinformation and deals with counterproductive defense mechanisms.

In response to this teacher and others like her, the museum and the schools should actively educate people about cultural and historical differences. On the one hand, we cannot pretend that America is not part of a long history of religious and cultural intolerance. On the other hand, the museum has to resist the temptation of becoming a site for the recognition of universal suffering and victimhood. The specificity of the Holocaust thus should work against false universalizations at the same time that it prevents any type of easy identification between the victims and Jewish culture as a whole. For example, the museum could undermine the equation of Jew equals

Holocaust victim by having a section on collective Jewish resistances to oppression. A more diverse depiction of Jewish identities could allow the museum to undermine the easy absorption of this cultural event into the universalizing and identifying structures that enable the easy assimilation of new knowledge into older prejudices and stereotypes.

Connected to this need to actively de-universalize the museum is a need to rehumanize the victims and the viewers. Like Gourevitch, I was taken aback to hear many young people describe the museum as being "cool," "exciting," "neat," and "awesome." Gourevitch further writes in his account that while he was standing by the shielded video monitors depicting the horrific experiments Nazis performed on their Jewish prisoners, he heard a young boy exclaim: "Pretty neat, huh. I mean, really sick." (I myself witnessed a boy's similar reaction on another occasion.) This self-conflicting rhetoric may point to how this student felt fascinated and excited by something and then realized he was supposed to be horrified by it. After all, why shouldn't he react to these violent scenes in the same excited way that he reacts to an action movie or a destructive video game? The fact that this horrible event is a historical reality is overcome by the way that it is presented on a video monitor to a student who has been brought up watching fake violence on television and in video games. In other words, we cannot expect people to turn off their usual viewing habits and processes of assimilation just because they are in a museum. Furthermore, Jean Piaget's theory of assimilation tells us that people will internalize new information by subjecting all new knowledge to old scripts and categories, and thus all educational efforts must find a way to reveal and then counter the previous representations that block learning.[7]

Psychoanalysis also helps us understand why people are attracted to the depiction of death and destruction. For, one of the central functions of universalization is to master traumatic material by subjecting it to symbolic repetition and control. By placing horrific scenes on video screens and on large framed pictures, the museum is able to provide a space for symbolic mastery; however, the excitement and enjoyment that many people express at the museum also point to the use of empathic identification as a way of making an experience seem real and authentic.

The effect of cultural cues and assimilation on the visitors is evident in the following student statement quoted by Gourevitch: "The pictures are disgusting—it wasn't a joke. But it seems like a long time ago because it was all black and white. It was a long time ago because, like, now we all get along together." If I am not mistaken, this statement of universalized tolerance repeats Rodney King's call for us to all just "get along with each other." Of course, King's video was in color, and it therefore seemed more real to the viewing audience who may equate black-and-white depictions to the old times before color television and the acceptance of different colors. Moreover, the distinction between now and then helps to separate the viewing audience from the horrors of the past and allows for a false universality of togetherness to reemerge. I would argue that in order to combat the effect of televisual distancing, the museum needs to rethink its reliance on video representations. Furthermore, the museum should explore the ways in which the Nazis themselves dehumanized their victims by documenting their deaths.[8]

In fact, one of the strangest experiences I had at the museum was when I realized that the killers themselves took many of the pictures of the helpless Jewish victims they killed. Did this mean that when I looked at these pictures I was looking from the perspective of the Nazis? Perhaps it is this unconscious identification between the visitors viewing the pictures and the Nazis shooting the pictures (and the people) that helps to explain the following statement that Gourevitch found in the visitor comment book: "This was great. We really enjoyed learning about all of the horrible things that happened in Nazi Germany." One could say that it was great that one learned something new at the museum; yet, the combination of enjoyment and horror may point to a vicarious form of enjoyment and perverse pleasure. While I do believe the museum is aware of this possible transformation of horror into fascination and viewing pleasure, I also think this problem has to be attacked head on. People need to be convinced that these videos are not just old black-and-white films and that one should not come to the museum with the same expectations of coming to a video arcade. In other terms, the museum has to actively reposition viewers away from their "normal" way of assimilating new cultural knowledge.

This does not mean that the museum should go out of its way to make people uncomfortable or to present things in a chaotic fashion, but our very notions of order and comfort have to be questioned and examined at the museum, and this type of educational and psychological intervention could be done through posted questions or live tour guides.[1] While the latter option would be costly, it would help work against quick acts of catharsis and identification. With live witnesses or teachers, the museum could rehumanize some of the abstract and mass-mediated representations that dominate the museum.

The Need for Dialogue

This need for live educators became very apparent to me when I was walking through the children's exhibit and heard a young boy ask his father: "Dad, what's a concentration camp? Is it a place where people go to concentrate, or is it a place where they kill people?" I was very curious to hear how this father would respond to his son's important question. The father paused for a second and replied: "It's a place where they put people who have been bad. Kind of like time-out." After hearing this man, I had to ask myself whether the father really believed this, or was he just trying to make things understandable for his child? In any case, someone needs to tell this boy that people were killed just because of their identity, not because of what they had done. But most of the voices that the museum presents are recorded voices that cannot respond and enter into a dialogue with the visitors.

After this father gave his son this incredible response, I saw his wife approach with a disturbed look on her face, and I thought that maybe she would add some wisdom to the family experience. However, she quickly let me down by stating: "Honey, we have to hurry up. I'm not sure what time the aquarium is open until." Maybe there is no difference for these tourists between seeing fish in tanks and Jews in camps. Of course, according to the false logic of the father, "the Jews" are in the camps because they did something wrong, but the fish, we just like to look at them.

Unfortunately, we also like to look at people suffering, especially if someone does not want us to look at them. If anyone doubts this point, just watch the children who gather around the video screens that are placed up high and out of reach of young viewers. One teenage boy near me shouted, "Move out of the way, I want to see too!" His friend replied, "It's really cool but it's not as good as the Auschwitz video." I later saw this same boy punching the same friend and calling him "Fag." I know the museum is not to be blamed for this behavior, but I do believe that the museum has to work with the preconceptions and prejudices that people bring into the museum. Moreover, this institution needs to actively work to educate people about the differences among the terms propaganda, historical facts, and fictional reconstructions. In fact, one of the more unfortunate aspects at the permanent exhibit is the placement of the film on anti-Semitism right after the windows on Nazi propaganda and race theory. When I was watching this film for the first time, a young man responded to the factual depiction of the history of anti-Semitism by saying to his friend, "Talk about propaganda." For this postmodern student, the Nazis have their propaganda and prejudices and "the Jews" have their own propaganda and prejudices. Thus, from the perspective of assimilation, all representations are social constructions that may or may not be true.

In response to this call for the museum to counter people's defenses and misconceptions, critics often posit that it is the responsibility of the parent or the teacher who is with these young people to educate them. However, it is clear to me that the adults are just as confused as the children are. In fact, when I was standing on line to see this film, two boys got behind me. An older man came by who was either their teacher or their parent, and he told them: "Come on guys. That's not important, we don't have much time. It doesn't matter anyway." I really wanted to ask this man *what* does not matter and where were they rushing off to, but I did not feel I was in a position to start such a dialogue. However, a dialogue is precisely what needs to be started inside and outside of the museum. Furthermore, this dialogue has to be structured by an understanding of how the various rhetorical defense mechanisms I have been elaborating on help structure the reception of Holocaust representations.

The Rhetorical Designing of the Holocaust Museum

In Linenthal's insightful analysis of the designing of the museum in Washington, DC, he reveals how the pedagogical desires of the museum were quickly countered by a variety of political and cultural concerns. In fact, throughout the lengthy development of the museum's design, we find a constant concern about presenting the truth of this traumatic event while not alienating a mass audience who may not like to encounter uncomfortable facts at a "tourist attraction." According to Linenthal, "The museum's mission was to teach people about the Holocaust and bring about civic transformation; yet, since the public had to *desire* to visit, the museum felt a need to balance between the tolerable and the intolerable" (198). The museum, in other words, had to be both entertaining and challenging in its combination of civic education and popular culture, and it is this merging of historical education and popular culture that serves as one of the basic issues in the popularization of the Holocaust and other cultural traumas.

In turning to Linenthal's descriptions of the museum's pedagogical and political rhetoric, I want to focus on the roles played by idealization, identification, assimilation, and universalization in the institution's main exhibit. In fact, the designers of the museum were acutely aware of the need to make this educational site both relevant to the average American and important to Jewish Americans (45). In other words, the museum was caught up in the more general social conflict between the desire to represent particular ethnic identities and the goal of representing a universalizing model of American citizenship. A central issue that was discussed in this debate over "postmodern" social assimilation was the decision whether to highlight the particular extermination of European Jews or to document the losses of the different identity groups who were murdered by the Nazis (55). Here we see how multiple, conflicting identity groups often shape the cultural realm of assimilation.

Another early question was where to place the museum, and the final decision to locate it on the Mall in the nation's capital brought up the issue of how this particular cultural trauma relates to American history and society. A connected issue, as Linenthal documents, was whether the memorial should serve primarily as a warning about the human propensity for evil, or should it be a place that preserves the human spirit and offer hope for humanity (110–11). This last question was clearly structured by the conflict between a rhetoric based on idealization and another rhetoric based on identification. As we shall see, the placement of the memorial/museum in the center of the American capital pushed designers to frame the depiction of the Holocaust as an essentially American event. However, every attempt to Americanize the Holocaust was matched by a strong desire to identify the Jewish people of Europe as the central victims.

Jewish victim identification was also combined with the desire to idealize America's role in saving Europe.

In fact, the museum skillfully works to position the viewers so that the audiences identify with the American liberators of the camps. Thus, in the initial elevator ride up to the first exhibit, the visitor is shown a video of U.S. soldiers encountering the camps. The viewer is therefore positioned by the museum to see things from the perspective of the idealized liberator who is represented by the shocked and angry General Dwight D. Eisenhower. In relation to this initial idealizing identification, Linenthal stresses how the designers were very aware that the memorial was on the Mall, and they felt they had to portray America in a positive light (101). However, we must ask, What is the effect on visitors when they begin this encounter with the tragic loss of life by being identified with the victim but seeing things from the position of the idealized American viewer? As I show throughout this book, popular cultural representations of the Holocaust often repeat this defensive gesture of idealizing the viewer by creating an identification with a liberator or savior who views from a safe distance the traumatic victimization of the "Other." What then helps make this depiction of the Holocaust tolerable for a mass audience is the production of an idealizing identification. That is to say, our ethical act or liberating gesture is confined to the mere act of watching other people suffer from an idealized position when we are viewers of trauma, and in the context of the museum on the Washington Mall, this encounter with trauma was used to shore up the idea of America as the great liberator.

Attached to this rhetoric of idealization we also find a strong rhetorical manipulation of identification by the designers of the museum. For instance, Linenthal argues that many members of the commission believed that " it was important to arouse strong emotion in the visitor in order to bring about a moral transformation" (112). Here we see how a pedagogical decision was made based on the idea that ethics are learned through empathic identification. Linenthal also points out that members of the commission wanted visitors to think about the problem of bystander indifference, and so the museum had to shake people out of their normal viewing habits (112). Thus, just as teachers interested in promoting positive social change in their classrooms may turn to traumatic identifications to wake their audiences out of their moral slumber, the museum creators were motivated to use victim identification as way of undermining a passive bystander mentality.

Linenthal reports that the designers of the permanent exhibit made a conscious effort to personalize the story and to give each visitor a way of individually identifying with the great loss of life (171). In this structure, the potential distancing caused by universalization and symbolization was countered by an emphasis on empathic identification. One way that the permanent exhibit tries to accomplish this task of empathic identification is by giving each visitor an identification card that has the picture, name, and story of a person who died in the Holocaust. The pedagogical idea behind this use of ID cards is to prevent the visitors from being passive bystanders by literally getting them involved in the story. However, as I discussed previously, one of the risks of this type of rhetoric and pedagogy is that it can create a fast and easy mode of empathic identification that appropriates the suffering of the other for private consumption. From this perspective, the assigning of fake ID cards to all the visitors may provide a false sense of identification; yet, this desire to get the audience involved in the "story" of the Holocaust is a central driving force behind many of the pedagogical and rhetorical decisions of the designers who now make popular representations of historical trauma.

In discussing this use of identification in the main exhibit of the Holocaust museum, Linenthal posits that the central way that emotion and identification are elicited throughout the museum is through the use of photographs of the victims. According to this logic, since photos often produce direct emotional reactions, the creators of the memorial thought these images would help to further personalize the exhibit. However, what is often missed from this educational strategy is an understanding of how emotional responses lead to ethical development. In fact, the assumption that images make the event come alive for the viewers fails to ask the question, What does it mean to simulate a direct experience of a stranger's suffering? Like reality TV and embedded reporters, this effect of the real may serve to create a quick emotional identification that goes nowhere and serves no higher purpose other than creating a sense of authenticity.

Perhaps the major problem with the design of the museum was the decision to see the site as a narrative complete with three major acts (168). I would argue that this symbolic structuring of the memorial helped the creators merge popular culture with modern education, but it also may have opened the door for a strong distancing and universalizing defense. As Linenthal states, "The permanent exhibit appears as a seamless tale, presenting its story through an anonymous voice that conceals those who shaped the exhibition" (168). Thus, by repressing the constructed nature of the museum, and by highlighting the narrative structure of the exhibit, the producers of this representation not only removed themselves and their choices from view but also fed the modern emphasis on unmediated, universal representation. Moreover, as I have been arguing, the constant use of media and narrative structures in the museum helps assimilate this educational effort into the popular cultural realm where fiction and fact are often combined and confused. Ultimately, the risk of this narrative move is a turn to history in order to combine it with fiction and narrative and simultaneously to create a safe space for forgetting and distancing.

To show how a different educational space can work against the possible negative effects of idealization, identification, assimilation, and universalization, I now want to discuss the Los Angeles Museum of Tolerance.[2]

The Critical Museum of Tolerance

While the Holocaust is the central topic of the Museum of Tolerance in LA, it also tries to connect this cultural trauma to diverse social issues. For example, the first exhibit that one encounters at this museum is a video installation that asks the visitors direct questions regarding their own prejudices and preconceptions. Furthermore, we find a map of "the Other America," which indicates the location of more than 250 hate groups in America. Thus, unlike the United States Holocaust Memorial Museum, the Museum of Tolerance begins by de-idealizing the viewer and America in order to actively engage all visitors and to work against the distancing processes of idealization and identification.

We can see this process of challenging idealization, identity, and subjective preconceptions as a psychoanalytic strategy based on the idea that the unconscious is essentially without identity and that one of the goals of analysis is to gain insight into how one develops a sense of identity through identification. Of course, many people will resist this "upsetting" process, yet psychoanalysis tells us that subjective and social change can happen only if people are willing to place their subjective investments and rhetorical defense mechanisms in play. In fact, I discuss different pedagogical processes later on that can be employed to aid this psychoanalytic approach to culture. For now, however, I want to examine how a *critical* museum can be an effective place for educating the public about traumatic issues and social prejudices.

One of the most compelling sites at the Museum of Tolerance is the exhibition examining the roles played by the media in the 1992 LA riots. Here we find an active attempt to get the viewers to think about the diverse modes of representation that are used inside and outside of the museum. One of the possible effects of this process is that it may challenge people to rethink the way they learn about history and culture through the media. This process of self-reflection also hinders the visitors' tendencies to see the museum as just another fictional representation. In the case of the Holocaust museum in Washington, DC, the only place where the systems of representation are called into question is in the section on Nazi propaganda. However, this analysis of culture becomes fixated on the Nazis and does not spread to an analysis of our own ways of representing history and culture.

In the case of the Museum of Tolerance, the question of representation is broached directly by showing how the museum itself was constructed and put together. For example, in the Holocaust section, several installations depict a historian and a museum designer discussing the evidence that they are going to present to the viewing audience. Here we see that the representation of history is always constructed out of a series of decisions made by experts and/or non-experts. This method of presentation denaturalizes the museum and helps open up a series of questions for the viewing audience. Moreover, by showing the ways that the museum and history are constructed, the installations provide a critical distance between the events depicted and the museum's narration of these events. In turn, this distance works against the quick idealizations, identifications, and universalizations that the audience may attempt to project onto the uncomplicated presentation of historical facts. Of course, this self-reflective display of the museum's own constructions may result in the subtle message that all history is fabricated, and thus the Holocaust may or may not have happened. At the museum in Los

Angeles, however, the critical model of historical reconstruction is coupled with a constant depiction—through slides and videos—of concrete historical events. Furthermore, the positioning of the Holocaust exhibit after the tolerance section helps shape the ways that viewers will perceive the narration of this historical trauma.

Unlike most other museums, the Museum of Tolerance sticks to an explicit method of education centered on the messages of the power of words, the role of individual choice, and the importance of personal and collective responsibility. These ethical issues are presented in the first part of the museum by constantly asking the visitors to question their own motives and prejudices. For instance, in the "Point of View Diner," visitors are shown a video depicting a scenario regarding prejudice, and then they are asked to respond to a series of questions by pushing in numbers on their personal jukebox. One of the effects of this installation is that it has people actively participate in the media representation of prejudice. By having people respond to questions that are posited on the video screen, the viewers are motivated to take a static object of one-way communication and turn it into a dialogical system of exchange.

In one of my class visits to the museum, we were asked to respond to a scenario concerning hate speech on the radio. One of the impressive aspects of this video was its attempt to depict all sides of an issue and to question the responsibility of all people involved, including the media itself. After we responded to these questions, the tour guide prompted us to have our own discussion regarding the issues with which we had just been presented. While some people might object that people go to a museum for entertainment and an encounter with interesting objects, not to receive didactic lessons, I argue that we need to see more interaction between the popular realm of entertainment and the critical area of education.

This desire to combine education, critical analysis, and popular culture can be seen as one of the driving forces behind a psychoanalytic and rhetorical version of cultural studies. One of the sources for this mode of analysis is derived from the way that Freud treated all symbolic creations, including dreams, as texts that need to be interpreted through dialogue and self-questioning. Furthermore, Freud's method stressed the idea of associating present experiences to past memories, and thus he took a historical approach to symbolic interpretation. While Freud tied all symbolism to shared cultural signs, he also posited that there is no single, universal language for dream symbolism. The inventor of psychoanalysis therefore constantly created dialectic among historicized culture, universal symbols, and individual meanings, and it is through this dialectic that a psychoanalytic version of cultural studies operates.

Just as Freud was constantly aware of the need to place unconscious symbolism into historical, cultural, and personal contexts, Aristotle's version of rhetoric provides an effective model integrating social, cultural, psychological, and political theories of representation. In fact, one can argue that Aristotle's rhetoric represents the first coherent theory of psychology, and this rhetorical theory is dedicated to a concern for how language works in a public arena. Thus, for Aristotle, every act of public speech must take into account the political position of the speaker, the social context of the speech act, the psychology of the audience, and the structure of the presentation. Rhetoric is positioned here to be an integrated theory of how language is shaped by psychology, politics, and cultural influences.

In combining Aristotle with Freud, we can see how the four main defense mechanisms I have been discussing need to be placed within a theoretical and practical framework combining history with sociology, psychology, linguistics, and politics. This interdisciplinary approach produces a psychoanalytic rhetoric that can work with both the necessary and the counterproductive roles that defense mechanisms play in education and culture. In fact, it is my contention that the Museum of Tolerance embodies this psychoanalytic rhetoric by constantly connecting the present to the past, the individual viewer to the general public, and the psychological to the social. Thus, by having installations that address the 1992 LA riots, the attempted genocide in Rwanda, and the Holocaust, the museum is effective in tying historical trauma and the media to current and past events. Furthermore, without using either idealization or universalization, the Museum of Tolerance opens up a dialogue among diverse groups and modes of representation.

An interesting way that multiple histories and multiple representations are depicted is in one of the first exhibits, where the visitor encounters a "man" who addresses the viewer directly and tries to bond with the audience by listing common prejudices and stereotypes. What I find most compelling about this installation is that the man is constructed out of video monitors, which are spaced in such a way that his body is fragmented. One of the effects of this fragmentation is that the viewer may begin the tour by identifying with a sense of multiplicity and a lack of unity. In fact, many of my students complained that they found the start of the museum to be disorienting. Therefore, as opposed to the United States Holocaust Memorial Museum, which starts with the viewer being identified with the idealized American liberators, the Museum of Tolerance begins with an active sense of de-idealization and fragmentation. In addition, this "video man" directs the visitors toward two doors: one that is marked for people with prejudices and the other for people without prejudices. Of course, only the door for prejudiced people opens, and so one is forced into taking on the role of being prejudiced. The critical message here is that everyone has prejudices, no one is ideal, and everyone is implicated in the topic of the museum.

One of the effects of connecting prejudice to multiple victims and multiple perpetrators at the Museum of Tolerance is the undermining of equating Jewish people, anti-Semitism, and the Holocaust while still maintaining the importance and specificity of Jewish suffering. For instance, the museum asks on several occasions why it is important to remember the Holocaust and other experiences of prejudice. The answer to this question is also formulated throughout the museum, and it points to the connection between multiple modes of past intolerances and current social, political, and personal formations of prejudice. The Museum of Tolerance thus de-idealizes the current visitor and American cultural history, while the Holocaust museum in Washington, DC, tends to idealize both the viewer and the country.

Perhaps the Museum of Tolerance is, in reality, a "Museum of Media Literacy and Prejudice," for its central contribution may be its ability to get a viewing public to examine how culture and history are rhetorically constructed and represented. This desire to transform the media productions of history into readable texts helps connect the circulation of prejudices to the domains of popular culture and public discourse. However, one of the possible problems with this procedure at the Museum of Tolerance is that it may serve to hide the differences and the uniqueness of the Holocaust, anti-Semitism, and multiple Jewish identities. Therefore, I believe it is necessary to supplement these sites of public discourse with a mode of rhetorical pedagogy that actively engages in the discussion and analysis of popular culture and the representation of diverse ethnic identities.

WHAT'S A CONCENTRATION CAMP, DAD? 51

By examining the Museum of Tolerance in relation to the United States Holocaust Memorial Museum, we see how the forces of idealization, identification, assimilation, and universalization often structure the question of ethnic identity in American culture. For example, the major narrative strategy of the latter is to contrast the identification with Jewish suffering in Europe with the idealized role of the American liberators. Here, the audience is motivated to identify with the great American liberators who save the powerless Jewish victims. In fact, the depiction of helpless Jewish suffering functions to idealize the American saviors by opposing the horrible fate of victims in Europe to the ideal state of American freedom. Of course, this narrative overlooks the fact that the Americans did not enter the war to save the Jews from destruction in Europe. While some mention is made of American neglect of Jewish suffering during the war, the museum's narrative is framed by the power of American liberation.

In contrast to this idealizing tendency of the museum in Washington, DC, the Museum of Tolerance tends to de-idealize America and the museum's visitors by pointing to the history of prejudice both inside and outside of the United States. In fact, the narrative of the Museum of Tolerance invokes a retrospective version of history by showing how current modes of prejudice lead up to the horrors of the Holocaust. From a certain perspective, this museum tells us that the universal prevalence of prejudice is proven by the particular experience of Jewish victimization. The Museum of Tolerance thus tries to get everyone involved in the story of the persecution of Jews in Europe.

CONSERVATIVE VS. LIBERAL MUSEUMS

The differences between these two museums can aid us in understanding the distinction between the "Conservative" and "Liberal" approaches to minority (and Jewish) issues in America.[3] According to this political logic, the conservative United States Holocaust Memorial Museum stresses the idealization of American democracy as a universal system that saves all minorities from their horrible pasts and gives everyone an equal chance at a new start. Meanwhile, the liberal Museum of Tolerance reverses this message and identifies the social and economic inequalities that minorities in America and elsewhere still face. In a sense, the museum in the nation's capital idealizes and universalizes America, while the museum in Los Angeles universalizes the nonideal state of ethnic minorities.

Of course, it is wrong to universalize all minorities and place them in one group. In fact, an interesting aspect of Jews in America is that most Jews do not see themselves as victims of economic and social inequality, nor do they identify with the idealized universal system that denies ethnic differences. A reason why many American Jews do not fit into either model is that their insecurities and real fears are not primarily economic or political in nature. Rather, the fears of many American Jews are derived from history, popular culture, and social interaction. In fact, I would argue that a major source of Jewish insecurity could be located in the recirculation of the most basic anti-Semitic stereotypes and myths in the popular media. Moreover, this type of popular cultural prejudice represents a new mode of discrimination in our culture, and throughout this book I concentrate on how internalized racism works in our current global systems of assimilation and mass media. I want to stress, here, that Jewish artists and academics themselves often circulate stereotypical representations of Jewish people in global popular culture. In other words, internalized self-hatred—a major aspect of the process of assimilation—has helped place Jewish people and other ethnic groups in a position where they profit socially and economically from their own self-victimization.

In the case of the United States Holocaust Memorial Museum, we can see the destructive effects of internalized anti-Semitism in the portrayal of helpless Jews that have to be saved by idealized American liberators and museum visitors. While many people would claim that this museum is still successful because so many people have gone to visit it, we must ask what have they learned, and in what ways does the museum play into the opposition between helpless Jewish victims and idealized American saviors? Interestingly, the Museum of Tolerance reverses this process of assimilated stereotypes by trying to use the example of Jewish suffering as a universal lesson about the effects of all forms of prejudice. While this lesson should be applauded, we must question what this structure does to the representation of Jewish people in America. Not only does it seem to lock Jews into the position of being eternal victims but it also poses the Jewish people as the source for the universal solution. By placing the Holocaust exhibit after the sections on prejudice and tolerance,

Jewish people are positioned to be the result of and the solution to the problem of prejudice.

What I believe both museums fail to do is represent multiple and varied aspects of ethnic and nonethnic identity. By centering Jewish identity on the Holocaust and anti-Semitism, these popular institutions work to identify ethnicity in mostly negative terms. Furthermore, as recent studies have shown—including Peter Novick's *The Holocaust in American Life*—American Jews themselves tend to define their Jewishness by their relation to the Holocaust and anti-Semitism (7). Perhaps these negative definitions of Jewish identity can be tied to the process of assimilation that motivates people to give up their particular cultural and historical identities in order to be accepted into the idealized and universalized dominant group. In many ways, American Jews and other ethnic minorities have allowed themselves to be identified by their oppressors, and thus Jean-Paul Sartre seems to be right when he declares that the Jew is invented by the anti-Semite. Yet, I would argue that Sartre could make this argument only because he also claims that Jewish people have no history, homeland, or culture. In many ways, by reducing Jews to their relation to anti-Semitism and the Holocaust, the Museum of Tolerance and the United States Holocaust Memorial Museum also participate in this denial of Jewish culture and history. To counter this reductive notion of assimilation and identity, I argue throughout this book that one of the most important roles for psychoanalytic cultural studies is to reverse this assimilating process by offering multiple and diverse avenues for identity and by critiquing the rhetorical defense mechanisms working to restrict identity formation.

THE PEDAGOGICAL IMPLICATIONS OF PSYCHOANALYTIC RHETORIC

By comparing the United States Holocaust Memorial Museum to the Museum of Tolerance, I have stressed the need for educators to take into account how popular audiences receive depictions of cultural trauma. In my articulation of the rhetoric of idealization, identification, assimilation, and universalization, I have started to outline the central defense mechanisms and reading strategies that educators need to consider when they seek to promote social change and critical thinking.

Moreover, I posit in this book that if these resistances to critical analysis are not dealt with in educational settings, most types of learning dedicated to promoting positive social change will be blocked. Thus, one of the pedagogical goals for educators teaching about the Holocaust and other traumatic subject matters should be the creation of a safe space where students are motivated to examine and work through their own modes of resistance.

I have found that one of the most effective ways of producing a change-oriented learning environment is to have students write short, nongraded personal response papers that can be later used in class discussions. The pedagogical idea behind these assignments is to motivate students to become active members in an interactive mode of education dedicated to respecting and challenging all student views. However, psychoanalytic theory tells us that this effort to make students open to thinking critically about their own ideas will most often generate anxiety since people, in general, resist learning any new information challenging their established views and identities. One thus needs to give students a chance to voice their own beliefs and resistances without fear of ridicule or being "marked down." However, I am not arguing here that teachers should simply reinforce and affirm students' opinions. In fact, I think that classes that stress the non-judgment of students' personal essays are often counterproductive: teachers need to be willing to allow their classes to become uncomfortable. The key is to manage the level of anxiety so that students are neither too comfortable nor too uncomfortable.

To both affirm and challenge students, it is essential to create learning situations where student resistances are recognized and analyzed. Importantly, psychoanalysis teaches us that resistance often fuels true learning and personal change. Resistances in the form of the four rhetorical models I have been discussing can both break and make an effective learning experience.

Psychoanalysis also tells us that one cannot address resistances directly and that before one gets people to deal with their defense mechanisms, one has to first create a safe environment and a certain level of trust. However, unlike other models of therapy, psychoanalysis attests to the importance of starting any type of critical self-analysis by bringing all modes of identity into question. While I do not think that education is or should be therapy, I do feel that many of the psychoanalytic strategies for provoking critical self-analysis can be employed to promote active student learning. For instance, I usually start my classes by having students write a short family history that asks them to respond to the following questions: (1) Where did your family come from? (2) What obstacles did your ancestors face in assimilating into American culture? and (3) How does tradition shape your present identity?[4] One of the goals behind this assignment is to open up a space in which students can consider how their identities are shaped by larger cultural and historical factors. I also want students to begin the process of examining the different aspects of assimilation and selfhood. It is equally important to use this assignment to allow students to recognize that everyone in the United States, except Native Americans, came from somewhere else, and thus we all have a divided cultural history.

While I do read over each of these initial responses, I do not comment on them or grade them because I want to create a safe place for self-expression. In fact, I have found that even in large classes, teachers can employ short, nongraded response papers as a means to get students to feel that they matter and that their opinions will be taken into account. Moreover, I use these responses as a way of jump-starting class discussions because, I have noticed, many classrooms require no active participation of their students, and in this type of impersonal learning environment, knowledge becomes static, like an object in a museum. Furthermore, as I argued before, the main way that this museum-like situation of education can be transformed is through active participation and critical question posing. Teachers will never know what is blocking student learning if they do not provide a space for personal reactions.

One method I have used to motivate students to participate in a class is to base half of the grade on the quantity and quality of their participation. While some may see this method as a way of manipulating students and forcing them to get involved, I have found that in our educational culture, which often caters to passivity and rewards people through grades, it is necessary to use evaluative tools as extrinsic motivators. In fact, one technique I have employed to help turn the current obsession with grades into a more positive lesson is to openly discuss with students my own conflicts over the need to grade students. One result of this self-disclosure is that students often start to see me more as a fellow human being and less as a teaching machine. Furthermore by placing grades in a larger social and systemic context, students see how diverse institutional forces usually shape individual judgments.

While it may seem that the question of grades has moved us far away from the topic of museums, education, and the Holocaust, I believe all these topics are united by the need to make learning more interactive, personal, critical, and self-aware. In other terms, if we do not take into account the resistances of students to learning important critical lessons, we cannot hope to overcome the passivity and defensive reactions that render people indifferent to the suffering of others. Whether we like it or not, grades, which many students see as markers for personal identity, often play the key role of either motivating or blocking learning.

If I am correct in seeing that the resistances to learning are the key to learning, then it is necessary for educators who are committed to critical thinking and social change to address the defensive use of idealization, identification, assimilation, and universalization in all educational situations. However, any direct attempt to address students' defense mechanisms will most often trigger more defensive reactions. In order to move past this stalemate, we need to develop pedagogical strategies that engage students' resistances in a safe and productive way. In the following chapters, I argue that the study of the Holocaust through the analysis of popular movies can be an effective method of working through students' resistances. As with my study of museums, I posit that a good way to start analyzing our own defenses is to look at how other people defend against threatening stimuli. The Holocaust provides an important subject matter because this historical trauma can work to call into question some of our most basic beliefs concerning modern Western culture. In fact, I use the term *postmodern* in the rest of the work to signal the way the traumatic events of the Holocaust have threatened to undermine the modern belief in historical and individual progress.

For many today, the Holocaust shows that technology often works to dehumanize people and that even an advanced democracy like Germany is susceptible to being transformed into a state of barbarism. Postmodern culture is therefore post-Holocaust culture, and it is important to provide students with this type of cultural history because another common aspect of contemporary postmodern society is a lost sense of historical awareness and context. In fact, as we shall see in the following chapter, the undermining of the modern belief in linear historical progress by the traumatic events of the

Holocaust has paradoxically created a situation where events like the Holocaust are often taken out of context and de-historicized. In turn, the decontextualization of the Holocaust threatens to turn this historical trauma into an object of popular consumption.

CHAPTER 3

REMEMBERING TO FORGET:
SCHINDLER'S LIST, CRITICAL PEDAGOGY,
AND THE POPULAR HOLOCAUST

In this chapter I explore ways of teaching about popular culture and the Holocaust to help students confront their resistances to learning new identity-changing information. I posit here that the interactive use of popular movies about the Holocaust and other historical traumas can introduce students, in a nonthreatening way, to the central rhetorical defense mechanisms that often shape the construction and consumption of contemporary culture. My goal here is not to simply provide another critical analysis of a popular movie; rather, I seek to use my interpretation of Steven Spielberg's *Schindler's List* to highlight different pedagogical strategies teachers can employ to unblock student resistances to critical thinking and social change.

A PSYCHOANALYTIC APPROACH TO
POPULAR CULTURE AND PEDAGOGY

While *Schindler's List* has been represented as an effective educational tool that keeps the memory of the Holocaust alive, I argue that the film really functions by employing the rhetoric of idealization, identification, assimilation, and universalization to allow the audience to forget about identity-threatening traumatic material.

Furthermore, like all objects of popular culture, this movie is pedagogical, yet its own pedagogical processes are unconscious and uncontrolled.[1] One of the goals, then, for analyzing this type of film in an educational setting is to make the unconscious mechanisms of popular culture conscious for students. I will also emphasize that this psychoanalysis of popular culture requires the development of certain techniques and concepts, which transform the passive consumption of entertainment into an active interaction with cultural information.

The need to render the content of popular culture more conscious for students can be derived from Freud's incredible insight that most art functions by matching the unconscious materials of the artist with the unconscious desires and fears of the audience.[2] To be precise, psychoanalysis posits that artists are fundamentally unaware of the source and meaning of their own creations and that audiences are also unaware of the reasons why they respond to art in a certain way. A psychoanalytic approach to culture thus requires the process of making the unconscious material of the artist and the audience more conscious. While some may argue that popular culture is not art, I would respond that a central aspect of our contemporary culture is the loss of the difference between what has been traditionally known as high and low art.

Contemporary society combines traditional art with mass culture, and contemporary methods of producing culture rely on technologies that speed up the delivery of information. In fact, movies move so quickly and provide so much information and stimuli that it is often impossible for people to consciously analyze the material they are internalizing. Moreover, by sitting in the dark and focusing on the giant screen, viewers are motivated to suspend critical judgment and relax and enter into the picture. In other words, films simulate the dream state where one lies in darkness and focuses on a series of images that are presented to the viewer without the viewer's conscious control or choice. Furthermore, this movie-watching situation is also comparable to a hypnotic state wherein the focus on a bright light or on the voice of the hypnotist puts the patient's critical mind to sleep.

As an educator dedicated to promoting critical thinking and social change, I had to find ways to make films less unconscious and hypnotic. The central pedagogical mechanism I chose is to constantly stop the quick flow of information and ask students questions to probe the meaning of the material they are internalizing. Of course, this constant interruption bothers many students because they want to escape and get lost in the pure flow of the images. I believe, however, that many teachers make a great mistake by giving in to the students' desires and showing media in class without stopping and examining the material. Furthermore, while some teachers may object to this process, claiming that it is essential to see a film as a completed whole and to allow for the full development of the plot and the characters, I posit that the linear unfolding of the story in the quest for narrative closure often functions to protect movies from critical scrutiny.

Perhaps the greatest common resistance to critical thinking concerning popular culture is the idea that movies and other forms of entertainment are just for enjoyment and that they have no inherent value or meaning. Importantly, in the case of the presentation of the Holocaust in popular media, it is hard to accept the idea that the popular depiction of history is really about nothing. Rather, I posit that the rhetorical claim of the meaninglessness of culture is really a statement indicating that we often internalize media in a nonconscious way.

The rhetoric of idealization can similarly block critical analysis of popular culture. When students and the general audience idealize a particular work or director, they often find it hard to engage in a process of critical analysis. This problem is especially acute in the case of studying movies by Steven Spielberg, who many people unconsciously associate with idealized childhood memories of great entertainment. Furthermore, in his turn to the Holocaust, Spielberg himself rhetorically places his moviemaking in the context of a personal religious awakening, and here the idealization of movies becomes attached to the idealizing tendencies of religious commitments.[3]

IDEALIZING OSKAR/OSCAR

In both a subtle and a not-so-subtle way, one of my central pedagogical strategies is therefore to de-idealize popular culture while simultaneously showing that it is not just meaningless entertainment. This strategy of intervention derives in part from Jacques Lacan's notion that a psychoanalytic interpretation is supposed to stop the speaker's flow of discourse and introduce a radically new line of associations. Moreover, for Lacan, every interpretation should act to undermine the patient's idealization of the analyst as the one who is assumed to know the answer to all questions. If we now transfer this theory of interpretation to a pedagogical situation, we can argue that one of the only methods to get students to see things in a new way without fostering a relationship of dependency is for the teacher to intervene in a non-idealizing fashion.

In the case of teaching *Schindler's List*, I begin this de-idealizing strategy by arguing that this film is really about Steven Spielberg's unconscious quest for an Oscar. Thus, in order to get this Oscar, Spielberg makes a film about the good German Oskar Schindler, who not only is able to save more than a thousand Jews but also is able to save all of us from the traumatic horrors of the Holocaust. What saves us in part from our own connection to this historical trauma is the way that we are placed in a position of identifying with the idealized Nazi Savior, Oskar Schindler, who is also equated with the idealized director, Spielberg. In this structure, seeing is not only believing, it is also liberating and purifying. For, like the American soldiers who frame the narrative of the United States Holocaust Memorial Museum in Washington, DC, the one who witnesses the dark crimes of the "Other" is also the one who liberates the survivors and the dead from their horrific existence. According to this logic, the very act of watching a movie or going to a museum is equivalent to an ethical act of heroic proportions.

The film both insists that "the List is life" and also implies that to save one person is to save all humanity. But who, exactly, is being listed and saved by Spielberg? Doesn't Spielberg save himself in the same way that he saves his Oskar, namely, by transforming him into nothing less then a towering Jesus figure in the form of the Direktor of a factory?[4] Yes, Schindler and Spielberg are both directors who claim they never have received the respect and honor they deserve. As Schindler later posits, however, the thing he has always been lacking is now present—war. Likewise, Spielberg has received two Oscars for making war films. In other terms, Spielberg realized that what he needed in order to receive the idealizing rewards from Hollywood was to produce his own Oscar/Oskar in the form of an idealization of a German Nazi. After all, if a Nazi can be turned into a Jesus figure, then all forms of personal redemption are possible.[5]

One thing I try to teach my students is that this rhetorical process of purifying and idealizing the Nazi subject is performed through the superimposition of multiple levels of identification. For example, in the opening scenes, we see things from the perspective of Oskar, who himself is identified with being a direktor/director. In other terms, we, the watching audience, are equated with Schindler and Spielberg as the ones who direct the point of view and the flow of the narration. Moreover, in the second scene, the camera approaches the action from the position of Schindler's hand, in which he holds a large bill. This rhetorical connection between money and viewing helps connect Schindler to Spielberg's reputation of being a big money earner at the box office. Spielberg starts the film by stressing the direktor's/director's purely capitalistic motivations: like Spielberg, Schindler at first uses money to get what he wants, and what he wants is more money and recognition.

For many critics, Spielberg represents the stereotypical director/producer who exploits his vision for economic advancement.[6] In the many interviews he gave about this film, however, Spielberg argued that *Schindler's List* was not about making money; in fact, he thought he would lose money.[7] Nevertheless, by first connecting his own directorship with the capitalist Schindler, Spielberg unconsciously attempts a double act of purification. If the capitalistic Nazi German can be idealized, so can the director himself and, by substitution, the viewing audience. No wonder people love this film: it is presented as remembering one of the worst traumas of Western culture but may also function to redeem the Nazi and the capitalist at the same time.

When I try to show students how this rhetoric of idealization and identification is used in order to train them, the audience, to respond in a certain way, they often react nervously—by moving in their seats, putting their heads down, sighing. I believe this nervousness is a sign that they are starting to resist my interpretations. Whenever I feel that this anxiety may be affecting too many students, I stop my analysis and have students write about what they are feeling and experiencing. I tell them to try not to censor themselves, that they will hand in these exercises in an anonymous way. What I am looking for is simply some quick and honest feedback. This method of free writing or free association draws on the psychoanalytic idea that people can overcome their resistances only if they first gain access to unconscious material, and the best way to access the unconscious is to write or speak without thinking about what others may think. In other words, one must be able to write without fear of censorship, and the use of anonymity often functions to suspend the fear of the teacher's judgment.[8]

I do not pretend that I do not have authority in the classroom or that I do not have my own views. However, I also tell my students that I do not consider my interpretations to be the only interpretation, and I welcome alternative views. While I center my interpretations on an analysis of defense mechanisms in the production and consumption of texts, I also use these interventions to provide a starting point for the students' own free associations. While many teachers may likewise claim they are interested in clearing a space for students to work on their own responses to class material, their use of grades and tests usually transforms the teacher-student relationship into a situation where the students are motivated to just give back what they think the teachers want to hear. In fact, as Lacan argued so well, most forms of therapy work in the same way. Therefore, true analysis and education start only when the mirroring structure of everyday conversation is undermined and the speaker begins to free associate without being concerned with what the other is thinking. In a classroom situation, a careful use of nongraded and anonymous free writing experiments can help students write without thinking about pleasing their teachers, and this process can be a necessary first step to critical analysis and writing.[9]

I am not advocating a class based entirely on free writing, but what I do want to argue for here is a mode of pedagogy that starts with free association. For example, once I have students do a quick five-minute free write, I collect their responses and I tell them that, at a later date, I will discuss the various trends I see in their emotional responses to the movie and my teaching. When I tell them this, I make sure to remind them that I will not be grading them on their responses because I will not know who wrote which response. In fact, I try to collect enough responses from enough different classes so that I can introduce reactions from other classes. I discuss some of these responses later on, but for now I want to return to my critical reading of the rhetoric used to structure Spielberg's film.

Rethinking Assimilation

One of the most difficult aspects for students and academic critics to accept is the idea that Spielberg unconsciously recirculates and assimilates some of the most basic anti-Semitic stereotypes and prejudices.[10] For example, when Schindler needs to get money to start his business, he turns to "the Jews" who have already been placed in the Krakow Ghetto. In many ways, this scene reinforces the anti-Semitic Nazi fantasy that "the Jews" controlled the banks in Germany before and during the war. In other words, even after the Jewish population has been displaced and oppressed, the Germans still have to go to them to get their money and labor. Furthermore, an early scene depicts "the Jews" in a church conducting black market business. The question that can be posed to students here is whether or not Spielberg is unconsciously repeating the classic anti-Semitic stereotype and myth that "the Jews" are destroying Christianity and Germany through their amoral capitalistic endeavors.

I try to get my students to consider the possibility that Spielberg's unconscious assimilation of anti-Semitic constructions points to how postmodern subjects internalize and perform the stereotypes that oppress them. In Spielberg's case, this mode of prejudice is partly because of his desire to reach a mass audience that may recognize only the most exaggerated representations of characters. By playing on stereotypes concerning Jews, Spielberg thus renders these characters recognizable and larger than life. For example, the Itzhak Stern character can be read as the embodiment of the stereotypical Jewish accountant who is small in stature and will help bend the rules to serve his master. In fact, many critics have pointed out that most of the Jews in the film are small and nebbish, while the two main German leads are tall men of great stature (Hansen, 83; Bartov, 49). Moreover, these small "Jews" spend the first half of the film talking about virtually nothing but money and their possessions.

In a very telling scene, "the Jews" are made to remove the gold from teeth that have been extracted from Jewish prisoners. This scene is a historical re-creation of an actual event, but what is strange is the movement of this gold: at the end of the film, "the Jews" turn gold into a ring that is then given to the great Nazi Savior Oskar Schindler.[11] Furthermore, can we not see the final resting point of this gold in the Oscar statue that Spielberg himself will receive? Thus, something is taken from "the Jews" (their memory) and is turned into a symbol that idealized the saving Gentile Nazi and the moneymaking director.

To help my students consider ways that assimilation motivates people to use and internalize self-degrading stereotypes and prejudices, I examine with them the fact that so many comedians and entertainers are minorities who make fun of their own ethnic groups. Furthermore, I ask students to think about how rap songs often represent the worst stereotypes that non-Blacks have about Black people and that television talk shows often get disempowered people to present stereotypes about their own identity groups in order to become popular. Finally, I ask students to think about how their own high schools were divided by different identity groups who all tried to fit in and be popular by copying certain positive and negative stereotypes. This final pedagogical move is directed toward arguing that the word *popular* in the phrase "popular culture" relates to how people try to be popular by fitting in and assimilating self-destructive stereotypes.

A central aspect of popularity and popular culture is sexuality, of course, which is often a difficult subject matter to discuss in a college classroom, especially in a small class that requires personal responses. Yet, psychoanalysis and everyday experience tell us that sexuality is essential to identity and subjectivity, and therefore any class really interested in having students explore their identities and defense mechanisms cannot shy away from this topic.

One way that I bring in the relationship between sexuality and identity is by showing how films like *Schindler's List* use eroticism in order to manipulate audiences and to block critical analysis. Thus, in Spielberg's movie one finds a rampant use of sexism and a perverse eroticization of suffering.[12] For example, in one scene the "bad" Nazi Amon Goeth begins to make sexual advances toward his Jewish maid. As other critics have pointed out, this scene of seduction is heightened by the Jewish character's attractiveness and her wet transparent shirt. In analyzing this scene, I have my students consider the question of whether Spielberg utilizes this moment of erotic titillation to seduce his audience or to try to explain the psychology of Nazi perversion as so many other films (*The Night Porter* and *Cabaret*) have done. Perhaps an answer to this question can be found in the shower scene, for critics have pointed out that the women being sent to the shower are depicted in a highly erotic way (Horowitz, 128; Loshitzky, 111). In fact, we first see the insides of the showers through what can only be called a "peephole." Moreover, the women are mostly young and attractive actresses who do in fact get wet for our entertainment.

Perhaps the most horrible aspect of this shower scene is the way it repeats a classic argument of Holocaust deniers: namely, there is very little proof that the showers were a killing machine and not just a hygienic mechanism. The absurdity of this claim, however, is clandestinely supported in the film during two key moments. In one of the first scenes at the concentration camp, a woman is discussing with other prisoners a rumor she heard concerning the way that showers were being used as a mode of mass murder. The other women in her room tell her to stop telling horrible stories. These stories were indeed true, but the film depicts them to be rumors or false representations. In another key moment, when the women do enter the shower in the later scene, all that happens is that they do get wet.

Once again, the question remains, Why would Spielberg not only repeat anti-Semitic stereotypes but also the discourse of the Holocaust deniers? Although some may say he is trapped by a novel that he did not write, that novel is very different in its depiction of these particular scenes.[13] Moreover, the novel itself is a creative reconstruction made mostly out of sources that may or may not be historically accurate. Consequently, the film is a pseudo-documentary of a fictionalized history, and thus it is presented through several layers of representational reconstruction. Furthermore, the film obsessively cites and repeats scenes and images from such past films as *Citizen Kane*, *Shoah*, *Cabaret*, *Night and Fog*, and *The Sorrow and the Pity*. This postmodern recycling of past representations of the past points to a general problem of history and memory in our current popular culture.[14] As in the example of using women in the film solely for their recognizable attractiveness, Spielberg also uses stereotypes and stock images to cater to a passive audience that wants to be entertained, not enlightened.

The Pedagogy of Empathic Identifications

One of the main points I make to students about this rhetoric of assimilation is the idea that the film may use memory and history in order to remember to forget about the traumatic parts of our past.[15]

In response to this position, people often reply that the film does get the audience to feel sympathy toward the Jews and anger toward the Nazis. However, we must ask ourselves what we learn through this quick process of cathartic identification. By becoming emotionally involved in this "history," do we just remember to remember but forget what we are supposed to be remembering?[16]

I would argue that this cathartic process often serves to block our ability to analyze critically the history that is depicted. In other terms, we laughed and we cried, but what did we laugh and cry at? And how will this laughing and crying help us deal with the horrors and ethical lessons of the Holocaust? Moreover, what kind of identification is produced in relation to the suffering of past victims? For, the reenactment of suffering may serve to place both the viewing subject and the viewed object in a position of being framed or frozen in time. For example, when one looks at the pictures of Holocaust victims taken before they are killed, one can become as helpless and as hopeless as the subjects who know that they are destined to die. On this level of the rhetoric of empathic identification, the viewer in front of the horrific object takes on the passive immobility of the object itself. In other words, one gets trapped inside the picture and becomes fixated oneself.

Furthermore, who has taken the pictures of the victims before they are put to death? Is it not the perpetrators of the crime who are also holding the machines of representation? We know, for example, that up to the time of the "Final Solution" the Nazis were obsessed with documenting the deaths of their victims. In fact, the killers themselves took many of the pictures that are displayed at Holocaust museums. There is thus an ethical problem that is rarely dealt with in relation to this collusion between documentary representation and murder.[17]

Universalization and the Death Drive

One lesson that students may be able to draw from this connection between death and representation is the idea that cultural reproduction does distance us from the reality of human suffering and existence. Thus, while students may resist embracing the Hegelian notion that through the rhetoric of universalization "the Word is the death of the Thing," they may be willing to consider how different modes of representation can dehumanize people and create a sense of symbolic control and mastery. Moreover, by using the Holocaust as an extreme example of technological and cultural alienation, students can begin to wonder about how they rely on technology and media as distancing devices. It is also important to ask students to consider how education is itself often a mode of using media and technology to distance people from real experiences.

In a strange and unconscious manner, Spielberg's film touches on this rhetorical connection between representation and death by making the "evil" Nazi Amon Goeth a double of the "good" Nazi Oskar Schindler. In one particular scene, this doubled identification is depicted by showing each of the characters as he shaves in front of a mirror at the same time. By cutting back and forth between Goeth and Schindler each facing the mirror of narcissistic self-reflection, Spielberg oddly equates our hero with our enemy. In fact, the second half of the film replays many aspects of the first half, with Goeth now playing the role of Schindler. Thus, we watch as Goeth picks his attractive secretary and shows himself to be a womanizer and a man of style and aesthetic taste just like Schindler.

In the middle sections of the film, Goeth takes on Schindler's idealized viewer role as the one who watches from above and determines the audience's and the director's point of view. Of course, Goeth is also a Direktor/director, and he is also in charge of exploiting "the Jews" for personal profit. Yet, unlike Schindler, Goeth is depicted as a hateful sadist who seems to pick his victims in an arbitrary fashion. It is important here to have students free write on why Spielberg would push himself, our hero, and our own viewing selves into the position of identifying with the man who always shoots after he gets someone under his sight? Is he equating the act of filmic representation with the process of not only shooting a scene but also shooting people?

I do not think we can credit Spielberg with having had such a profound insight as to link murder and his own processes of representation. What I do think is going on is a universalized model of subjectivity and identification that serves to equate the victims, the perpetrators, and the bystanders. Therefore, according to the logic, we are all innocent victims and evil perpetrators in a postmodern world that tends to universalize acts of suffering. Like great directors and writers, we must be able to empathize with all characters and emotions. In our obsessive desire to reconstruct the past, we want to act out the traumatic memories of the Other as a present experience. I would argue that, safe in our museums, living rooms, and movie theaters, we pay for scenes of pain and redemption that ultimately place us in chairs of total passivity. Like Schindler sitting on his white horse on a hillside high above the destruction of the Krakow Ghetto, we are impassive because we have but a distant vision of the Other's pain. Or perhaps we are like Schindler when he teaches Goeth how to pardon himself by pardoning others. Better yet, maybe we are more like Spielberg, who gets the attention and recognition he has always wanted by getting his Oskar/Oscar.

In a very telling scene, Stern brings a man without an arm to Schindler in order to thank the Direktor for employing and saving him. After this one-armed man blesses Schindler, the Direktor scolds Stern for putting a disfigured man in his presence. Schindler, like Spielberg, wants to see beautiful people, even in this unbeautiful setting, because beauty serves as a shield to hide the true essence of human tragedy. This covering aspect of beauty is revealed at several stages of the film and is most acutely evident in the scenes following Goeth's brutal actions. After several scenes of arbitrary violence, for instance, Spielberg cuts to the visual depiction of a naked female. Similarly, at another point, Schindler tells Goeth's beautiful Jewish maid, "He won't hurt you because he enjoys you so much." It is thus the enjoyment of constructed beauty that serves to save this subject's life. However, beauty blinds us to the pain of the Other, while it helps to dehumanize and objectify the subject of representation.

Popular Resistances

Once, after I had delivered a version of this interpretation of *Schindler's List* at a conference, an older woman stood up, and I could see that she was upset. I feared that she was a Holocaust survivor and was bothered by my critique of the film. To my surprise, however, this woman criticized me for undermining the greatness of Spielberg, exclaiming: "How could you put down such a beautiful film by such a great director? He has brought the Holocaust to so many people who may know nothing about it. Some things should just be left alone and respected."

This woman's response is a very typical one that people have to any criticism directed toward popular culture representations of the Holocaust. On one level, this defensive reaction is dependent on an idealization of the great author/director. Here we find a secular celebration of creative genius that may represent a displaced deification of the great individual. In this context, Spielberg is seen as a Jewish god or Messiah who has brought the holy message to his people.

On another level, this woman's comments point to a universalistic logic that argues it does not matter what is actually learned by this media representation, what matters is that it is represented for everyone. Furthermore, her call to respect the film points to the idea that the Holocaust is actually an idealized holy event and so must be treated as if it were an act of divine intervention and its representation a holy text.

I have encountered these same types of resistance to interpreting and thinking critically about this film among university students. Here are some of the reactions they have expressed after watching *Schindler's List*.[18]

> I do not think that we should analyze a movie that is so important and moving.
> It made me value my own life.
> It showed that even in bad situations, there are good people.
> It showed that in order to survive you have to keep the faith.
> I think it was very anti-German.
> It made the Holocaust seem very real.
> I do not know why more Jews did not try to escape.
> It could have been more balanced.
> It is good to see that some Germans tried to help out.
> Being in black and white made it seem more real.
> Spielberg really knows how to tell a good story.

Although these statements could depress a teacher, it is nonetheless important to analyze them in an open and thoughtful way because these types of resistance often block the potential learning experience that one can achieve through the critical analysis of popular culture. Furthermore, these rhetorical resistances are also common unconscious interpretive strategies. Therefore, in the pedagogical strategy I am endorsing here, the teacher does not simply dismiss these rhetorical responses but recognizes them as valuable yet limited methods of identity construction and learning.

The same rhetorical defense mechanisms I have identified as occurring in the film are repeated in the student responses. Some of this resistance revolves around the idea that any popular movie made by a popular director should be idealized and not analyzed. Here, the "art for art's sake" mind-set serves to privilege the idealized author over the content. In this context, students often defend against interpreting movies and books because we can never know the true intention of the author, and thus we can never really know what something is about. This viewpoint plays on the idea of an omniscient and absent god/author who should not be questioned or examined.[19]

Related to this idealization of the author/director/god-figure is the celebration of redemption and salvation. Two statements in particular—"It showed that even in bad situations, there are good people" and "It showed that in order to survive you have to keep the faith"—display a displaced notion of Christian redemption in a situation where redemption and salvation were virtually impossible. In fact, due to their lack of knowledge about the arbitrary and ruthless power of the Nazis in the concentration camps, students often criticize Jews for not trying harder and not keeping their faith. One reason for these unrealistic expectations is that students have gained most of their knowledge about the Holocaust through films and television shows, and these cultural productions, in order to retain a popular audience, most often provide an idealized happy ending and a message of redemption. Furthermore, these internalized and idealizing messages become the source for the popular audience's sense of historical truth. Someone teaching about the Holocaust thus must first expose and analyze the different false conceptions students bring to their encounter with this subject matter.

While some teachers may not want to openly criticize the misconceptions their students share in class, it is hard to imagine how education can create social and personal change if it does not challenge students' preconceptions and defenses. A key to this process is not to let the class atmosphere become so tense that all rational discussion is undermined (even though a certain level of anxiety and insecurity is needed to keep the discourse of discovery open). Once again, nongraded, anonymous free writes can be used to create a space where students can acknowledge and work through their rhetorical resistance. I also want to emphasize that even in large lecture classes, this type of free writing can be used since the teacher will not be grading or responding directly to the work.

Free writing assignments often uncover the common rhetoric of universalization. For instance, in contemporary college and university classrooms, one of the major false conceptions blocking the critical analysis of popular culture is the idea that every representation should be balanced. Two student responses especially—"I think it was very anti-German" and "It could have been more balanced"—reveal this tendency toward universal acceptance of all people and all viewpoints. Since our culture in general and our education system in particular preach the rhetoric of the mutual tolerance of all arguments and all ideas, teaching about the Holocaust creates a difficult situation in which the teacher and the students are forced to make ethical decisions regarding historical intolerances. Since students are so used to watching television talk shows and news programs where different viewpoints are represented and given airtime, they often shy away from taking a critical stand against any idea or person.

This rhetoric of universal tolerance is often coupled with its opposite: intolerance. Thus, on one level, students will claim that there are two sides to every story, and on another level, they will idealize the good guy and demonize the bad guy. I posit that in this psychological structure, universal tolerance is usually attached to a strong sense of idealized self-righteousness, and therefore educators who seek social change must work against universalized tolerance, self-idealization, and the demonization of the other. It is also essential to help students understand how different rhetorical defense mechanisms reinforce each other. For instance, the student comment "I do not think that we should analyze a movie that is so important and moving" reveals that the move toward empathic identification and idealization is coupled with a universalized call for cultural nonmeaning.

As I discuss throughout this book, a fundamental driving force behind the universalization and globalization of culture is the idea that American popular culture should not be analyzed and, in fact, that it has no inherent meaning or value. Thus, what may allow our popular culture the ability to move around the world is that its lack of perceived meaning prevents it from conflicting with any established ideology or belief system. By turning to popular representations of the Holocaust, I believe I have been able to convince some students that culture must have a meaning because it would be an act of Holocaust denial to claim that our representations of this cultural trauma are meaningless.

Integrating the Rhetorical Structure of Defense Mechanisms

To help clarify the roles played by the rhetoric of idealization, identification, assimilation, and universalization in *Schindler's List* and in contemporary popular culture and pedagogy, I have found it necessary both to integrate and to differentiate these distinct defense mechanisms. One method I use in my classes to help students conceptualize how these different mechanisms relate to one another is to present the following logical square:[20]

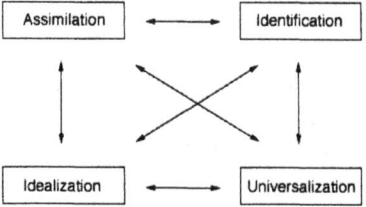

On the most basic level, this diagram posits that while empathic identification tends to replace the experience of others with the subjectivity of the viewer, idealization replaces the subjectivity of the audience with the power of the idealized author or character. In this sense, empathic identification is opposed to idealization, yet they are often combined. Likewise, there is an inherent opposition between the rhetoric of universalization and the structure of assimilation. While universalization erases all ethnic and cultural differences in order to promote the universal equality of the empty subject, assimilation is based on the reiteration of social stereotypes and essentialized ethnic differences.

If we apply this diagram to my reading of *Schindler's List*, we find the following:

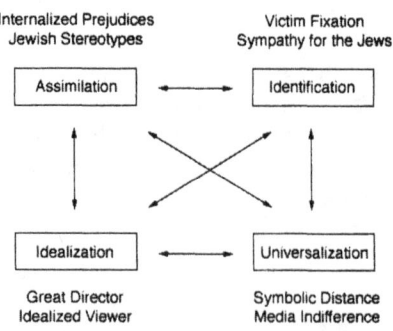

This integrated rhetorical structure posits that we need to understand how these four very different defense mechanisms work together. For example, the idealization of Schindler and Spielberg is dependent on the assimilation and recirculation of anti-Semitic stereotypes that not only essentialize Jewish differences but also feed into a victim identity. This, in turn, allows the audience to use empathy to identify with Jewish victimhood and to absorb the Other into the self. Feeding this structure of identification is the power of visual media to represent multiple subject positions in a state of neutrality or indifference. In other terms, because we believe that visual media have no inherent value or meaning, they are able to create a sense of symbolic distance and the illusion of viewer control.

We can also add to my rhetorical reading of *Schindler's List* the four central models of audience resistance to analyzing the meaning and value of popular culture:

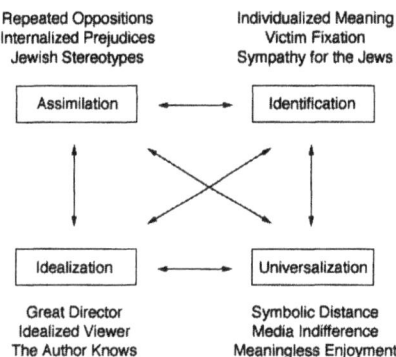

This diagram emphasizes how the universalized claim that media provide just meaningless enjoyment is countered by the way popular culture relies on the recognition and repetition of assimilated stereotypes and oppositions. Moreover, the idea that only the idealized author of a text can tell us what it is about is contradicted by the notion that every individual relates to a cultural production in his or her own way.

After mapping out these different responses, it is also important to place the four rhetorical reactions within the context of academic criticism. In fact, I have found that these four major rhetorical defense mechanisms directly reflect the four major theories of literary and cultural interpretation:

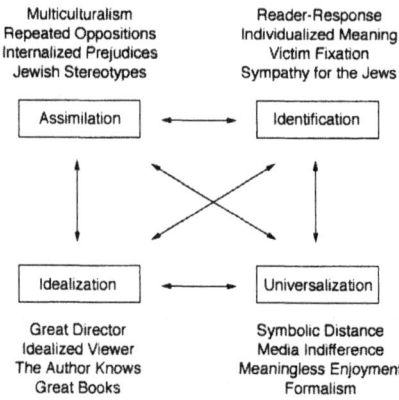

I am positing here that student responses not only match the major movements of literary criticism but also reflect on a historical and social transformation. One of the most effective ways of introducing students to critical theory is to show them how academics have employed different reading strategies to interpret works of culture and how these different strategies are shaped by different rhetorical defense mechanisms.

Thus, the Great Books theory of literature and culture argues that the only important thing to know about a work is the intentions of the idealized author. Here, we find a displaced religious conception of the author as a god who controls the universe of the text. Furthermore, the defenders of the Great Books tradition believe we should continue to teach these works because they reflect back on readers and allow them to encounter the best of what has been thought or said. In other terms, the idealization of the Great Books is supposed to result in the idealization of the reader who is enlightened by discovering the author's true intentions.

Yet, because most authors are either dead or absent from the scene of interpretation, the Great Books tradition has to be supplemented by other rhetorical approaches. The formalist strategy is one of the first major methods of interpretation both to help and to undermine the Great Books theory. Starting with what is often called "New Criticism," cultural formalists concentrate on detecting the inner laws that structure a particular work. Here, instead of being concerned with the intentions of the author or the biographical context of the writer, the critics turn their attention to the scientific process of discovering universalized underlying laws and symbolic systems. However, in this effort to locate the structure of a cultural production and to avoid its social, cultural, and personal context, formalists tend to promote a new type of aesthetic retreat into meaningless enjoyment. In fact, I would argue that what is often called deconstruction or post-structuralism in literary theory is usually another mode of formalism dedicated to the pure act of interpretation within the laws and structure of a particular text.

When students claim that films are just meaningless escape and that we should not try to analyze these works, they are often unconsciously repeating some of the same arguments of the formalists who do not want to attach texts to any particular social meaning or value. In opposition to this approach, we find the multicultural and postmodern responses centered on the notion that all works are shaped by important social and cultural forces. Moreover, these contemporary approaches challenge the scientific and cultural neutrality of the universalizing formalists by emphasizing the role of competing cultures and discourses in the shaping of all cultural productions. One possible problem with the multicultural approach, however, is that it tends to essentialize and assimilate social stereotypes in its efforts to argue for the importance of particular ethnic identities. Furthermore, many people feel that multicultural theories are based on a model of social determinism that does not allow for individual responses and agency.

We can therefore understand the reader-response theory of literary interpretation as a reaction to the social determinism of postmodern multiculturalism. From this ethnographic perspective, all readers are free to determine their own interpretations; thus *me*aning, appropriately, really begins with me. In turn, by placing the locus of the creation of meaning in the heads of the individual audience members, reader-response theories also cater to models of empathic identification. Here, we see that empathy allows the interpreter of a text the ability to absorb the experience of others into the privatized self.

One reason why I want my students to understand and think about these different theories of cultural reception is that I hope they see how the rhetorical defense mechanisms they encounter in others and in their own selves have a long shared social history. I have found it valuable to historicize this model by attaching the idealization of tradition and the Great Books to premodern society and the universalization of abstract forms to modernity. In turn, it is essential to help students understand the meaning and value of postmodernity by showing how we now live in a period where traditional premodern values and modern structures of universalization are being challenged by the assimilation of multiple and conflicting cultural interpretations. I also argue that we are entering into a social period dominated by a backlash against postmodern social determinism and multiculturalism, and this new order stresses the role of individuals and empathic identification in the creation and reception of culture. In fact, the recent rise of new media indicates the growing dominance of the "me" in *me*dia.

To help conceptualize this emerging social order, I have coined the term "automodernity," which is based on the idea that we are now witnessing a mutual stress on individual autonomy and technological automation. For example, in the rhetoric of empathic identification, we often see how new technologies and media allow people to maintain the illusion that they can actually simulate other people's experiences and that these simulations help them to empathize with the plight of others. Yet, as I have pointed out in relation to both the United States Holocaust Memorial Museum and *Schindler's List*, the technological simulation of the past may help people to falsely believe they are understanding or learning from history when they are simply using the past of others to access their own emotions and meanings.

The Holocaust and the Four Modes of Resistance

Another reason for my turning to popular representations of the Holocaust is to show how the processes of idealization, identification, assimilation, and universalization not only account for the ways that people view popular culture today but also how, in their most extreme forms, these rhetorical defense mechanisms structure the very nature of the Holocaust. For example, critics have often posited that the four central charters/roles of the Holocaust are the leaders, the collaborators, the victims, and the bystanders and that each one of these subjective positions is dominated by a certain psychological mechanism.[21] Thus, the leader idealizes his own role in history, while the followers idealize the person who leads them. In this rhetorical structure of idealization, the general populace gives up their own subjectivity in order to idealize the greatness of their leader, and here we see a regression to what Freud calls the primal identification of the primitive group. Furthermore, Freud helps us understand that in this idealized submission of the people to the leader, the powerful leader replaces the superego of the individual followers and allows for a regression to a primitive state where the instinctual id is released.

In turn, what helps define the collaborators is the notion that they have been placed into a bureaucratic system that shifts all responsibility onto the leaders. Thus, in Karl Adolf Eichmann's rhetorical claim that he was simply following his orders and doing his job, we see how the rhetoric of universalization accounts for a distancing from personal accountability. Furthermore, the rhetoric of universalization can also be located in the roles played by science and technology in dehumanizing the victims of the killing machine and turning human subjects into symbolic objects to be processed through modern assembly line systems. One can also argue that the failure of so many outside people and countries to intervene in this process indicates that people were either desensitized to the horror of the Holocaust because of the constant war and destruction proceeding World War II, or the particular plight of European Jews was lost in the universalized notion of "Total War."

Linked to this distancing of people from the reality of the Holocaust is the role played by German people who not only idealized Hitler and his ideology but also were taken in by Nazi propaganda. Central to this propaganda effort was the assimilation of anti-Semitism as a recognized European ideology. In fact, my theory of assimilation helps clarify an issue in Daniel Goldhagen's influential work on the cause of the Holocaust. According to his *Hitler's Willing Executioners: Ordinary Germans and the Holocaust*, the driving force behind the Holocaust is what he calls European eliminationist anti-Semitism. In this fatal form of prejudice, the long history of European prejudices against Jews is determined to be the only true cause for the Holocaust. While Goldhagen's thesis has many problems, the central one is surely his idea that the long history of European anti-Semitism somehow took on a much stronger and lethal form in Nazi

Germany. In contrast, I would argue that my four-part model is much more effective in accounting for the multiple causes for this historical event. For example, we must think about how the idealization of a leader was connected to the universalization of collaborators and the assimilation of a previously existing racist and anti-Semitic ideology.

We must also look at how the Germans identified with being the victims of previous wars and sanctions, and they used their own victim identification to justify their acts of revenge. Therefore, on the level of empathic identification, Hitler was exceedingly apt at manipulating the rhetoric of victim identity at the same time that he employed an idealization of himself and the German race. In contemporary culture and politics we often find this same paradoxical combination of victim identification and self-idealization. The idealization of aggressive policies is first attached to a victim status that removes the aggressor from any moral blame or responsibility. In this structure, to be an idealized victim is to be innocent and therefore beyond reproach.

Countering False Universalization in History

By attaching the common experiences of indifference, idealization, assimilation, and identification to the extreme horrors of the Holocaust, my pedagogical strategy is to provide students with an ethical and historical framework that may motivate them to take these rhetorical defense mechanisms more seriously. However, this process should not result in students proclaiming the universalizing position that there are no real differences between our culture and the culture of the Holocaust. Rather, historical comparisons need to stress the important differences between historical periods while simultaneously trying to make essential connections.

In the case of Nazi Germany, the idealization of Hitler and the other fascist leaders relied on the public's willingness to identify with the leader and the nation as an ideal entity. In many cases, the foundation of this idealization was tied to a displaced religious feeling of redemption and salvation. Although we still see this rhetoric of idealization in our own culture and politics, our popular culture often feeds off of the constant mocking of all authority. In fact, this undermining of all traditional sources for idealization can result in a devaluing of politics and other forms of social engagement. Moreover, what we often find in popular culture is that the only subject who is still idealized is the antisocial individual. From this perspective, one way that current audiences may assimilate the representation of strong figures like Hitler and Schindler is by seeing them as "great" individuals who have reshaped history. In this structure, the great leader is perceived as a role model for the great individual. Likewise, as viewers of an idealized film by an idealized director, we feel that we become ideal. At times, it does not matter whether this ideal figure represents radical good (Schindler) or radical evil (Goeth). What matters for the viewing audience is the pure power of the individual to shape history and society. Furthermore, these ideal figures often stand above history and judgment, and thus they feed into the desire of the audience member to be a purified spectator of culture.

An exploration of the rhetorical defense mechanisms utilized in Nazi Germany also teaches students the lesson that in order for the idealized leader to maintain power, he needs collaborators and people who are willing to follow his orders without judgment. As I stated previously, many of the Germans who were convicted of war crimes pleaded that they were only following orders, and thus they could not be blamed. However, what is often ignored in this discussion is the way that our own current culture also asks people to repeat symbolic orders and rituals without questioning them. Thus, one of the defining aspects of any bureaucratic system is the way that it combines symbolic knowledge and practices with a displacement of responsibility. For instance, in many contemporary educational settings, this same type of bureaucratic distancing is apparent in the ways that students are trained just to learn information so they can perform successfully on standardized tests. What we often find here is a social structure in which only the expert (the teacher) is allowed to ask questions, while the audience (the students) must internalize symbolic knowledge without actively engaging in analyzing the new information. Not only does this structure tend to idealize the teacher as the one who knows but also to transfer the students' responsibility to think critically onto the teacher, who controls the flow of discourse.

In order to break out of this type of educational structure, I advocate that education and culture need to be made more interactive and intersubjective. One way to help convince students of this need for dialogue is to show them the extreme situation in Nazi Germany where culture, education, and politics combined idealization, identification, assimilation, and universalization in a noncritical way. For instance, people did things they might not have done if they had thought about the consequence or meaning. However, Hitler's rule allowed for a denial of personal responsibility and concurrently produced an overly abstract and generalized view of human beings.[22] It is precisely this abstraction and assimilation of generalized stereotypes that we need to combat in the study of popular culture. The tendency to simply say "It's only a movie" or "It's just entertainment" hides the fact that these cultural productions are sending powerful messages. Moreover, as in the case of Nazi Germany, Hitler was able to seduce his public by circulating fictional prejudices that appealed to an uncritical audience.

Another aspect of the Nazi use of symbolic culture was the employment of modern media, transportation, and communication systems to unite the public and to destroy and dehumanize people. By combining science, technology, and an amoral bureaucracy, the German killing machine was able to universalize all aspects of modern culture and direct these new machines and techniques toward the industry of death. In fact, part of the dehumanization process that helped allow the Germans to treat their victims as pure objects was the perpetrators' ability to deny the Jewish people their cultural and historical traditions and definitions. By negating their cultures, the oppressors were able to define the victims. In many ways, this process is still going on today in popular culture when groups are defined by a founding trauma or state of victimhood. One could argue that by basing identity on an original scene of trauma, diverse ethnic groups may fall into the trap of allowing the oppressor to define them as a victim.

In the case of Nazi Germany, the leaders used the rhetorical assimilation of symbolic stereotypes to deny the victims their culture at the same time that the victims were given new identities. Thus, the use of yellow Jewish stars and numbered tattoos fixed a certain identity marker onto the body of the victim. In *Schindler's List*, we find a strange repetition of this process when we see the Jews being represented as inferior to the Germans. We also find this mode of identity making in the internalization of oppressive stereotypes and prejudices. By identifying with the aggressor, victims are able to participate in an unconscious fantasy wherein power is derived through identification. Likewise, the audience watching scenes of torture and abuse feel empowered by identifying with both the perpetrator and the victim. In this fantasy structure, the identification with the oppressor enables the audience to feel a sense of pure power, while the identification with the victim gives the audience a concrete sense of identity and objectivity. For, what often fascinates people about the Holocaust is the sense of raw emotion and violence coupled with a clear sense of the difference between good and evil.

In analyzing with students these extreme uses of culture and symbolism, they can begin to gain a more critical understanding of how our own society uses the rhetoric of assimilation. For instance, the Germans not only relied on shoring up their own identity by having a clear enemy but also allowed for a free realm in which people could act out their most basic sexual and aggressive desires. In many ways, popular culture is our safe realm and haven for the expression of otherwise repressed instincts. Furthermore, our systems of popular culture often feed into a passive bystander type of mentality. Just like the millions of people who pretended to know nothing about the Holocaust, our culture tends to stress the innocent bystander point of view. For example, in the repeated claim that our culture is just meaningless enjoyment, and thus should not be analyzed, we find a universalized position of tolerance coupled with indifference. Moreover, in this universalized position of tolerance, all differences are overlooked and all moral judgments are rendered suspect. Like the views of many of my students, the official position of liberal tolerance often degenerates into moral relativism. However, the problem with this universalizing view becomes strikingly apparent when one is dealing with a subject matter like the Holocaust.

Yet, these comparisons between our own contemporary culture and Nazi Germany should not prevent us from losing sight of the major differences between these two cultural orders. In fact, we can clarify the meaning of our own popular culture by distinguishing it from the culture of the Holocaust, for we live in a period when most modes of traditional authority are constantly being undermined by larger global economic forces. We also have a strong tradition of divided governments and checks and balances that work against the consolidation of totalitarian power. Furthermore, the popular expressions of prejudice and anti-Semitism have moved from the realm of direct representation to a much more subtle and internalized form. In movies like *Schindler's List*, we see that the explicit state-sponsored racism of the modern Holocaust is transformed in the postmodern period into an implicit recycling of destructive stereotypes for the pleasure of the popular audience. In fact, Spielberg not only represents internalized anti-Semitic messages but also couples them with an idealization of the oppressor. Because these representations are, in this instance, contained in the realm of entertainment, they are not considered to be dangerous or harmful. However, I would argue that these popular culture productions are recirculating messages of hate and discrimination in an unconscious manner. Instead of having an organized system of state-sponsored hatred, we now have a decentralized industry of hatred under the banner of entertainment.

Psychoanalytic Pedagogy

One of the goals of a psychoanalytic rhetorical pedagogy is to provide tools that will enable people to interpret the subtle messages of hatred and prejudice that circulate in the popular media. To fight the internalization of anti-Semitism and other modes of intolerance, we must become more sensitive to the role that popular culture plays in the circulation of stereotypes and subtle forms of Holocaust denial. Moreover, we need to find ways of using education to counter the destructive employment of idealization, identification, assimilation, and universalization in popular culture and in the realms of public opinion and national politics.

To get students to work through the popular circulation of rhetorical resistance, I have them write research papers in which they must critique a popular film and study the roles played by the four modes of rhetorical resistance in both the construction of the film and the way people on the Web discuss it. I also ask students to examine the influence that these defense mechanisms have played in shaping one of their recent classes. A guiding idea behind this assignment is that students will be better able to analyze their own resistance after they learn to analyze the rhetorical resistance of others. A valid criticism of this assignment is that this process of examining the defense mechanisms of others will block a critical analysis of the students' own resistance. To answer that objection I discuss in later chapters how I get students to turn their critical attention to their own defense mechanisms once they have first gained experience in analyzing the rhetorical reactions of others.

In having my students locate the processes of idealization in both the production and consumption of media and education, I hope to show how rhetorical resistance is both necessary and limiting. I also want my students to see that education itself is a cultural product that needs to be analyzed in a critical fashion. Moreover, this critical research project ends by having students discuss the role of social and collective action in their chosen film. My idea here is that isolated individuals alone do not shape history and life; it is necessary to see how social change is tied to the participation of people in groups.

In fact, getting students to connect subjective rhetorical resistance to larger cultural forces can be presented in such a way that the individual and collective aspects of identity are both affirmed. For example, I am careful to point out to students the different places in *Schindler's List* that exclude how Schindler's activities were influenced and at times funded by Jewish groups. I quote passages from the book the movie is based on to show how Spielberg and his scriptwriters systematically shut out almost all aspects of collective Jewish resistance in order to stress the role of the great individual. This pedagogical process thus entails correcting the false impressions presented by the movie as well as displaying how American culture often uses the four main models of rhetorical resistance to base history and culture on the actions of an individual rather than on the collective actions of the group.

This need to emphasize collectivity in relation to subjectivity can also be revealed by asking students to examine the stress on individualism in their own educational environments. It has been my experience that students are quite willing and able to analyze their past classes in order to detect the conflict among idealized individualism, social assimilation, and globalized conformity. Therefore, by tying the critical analysis of popular culture to the analysis of education institutions, students are better able to see how these rhetorical defense mechanisms affect their own subjectivities on a daily basis.

This method of analyzing the defense mechanisms of others is only a first step—albeit a valuable one—and what I discuss in the next chapters are some of the pedagogical techniques I have used in order to bring the focus of the students' critical thinking back to their own lives and identities. Before this process of self-reflection can begin, though, it is often necessary for the teacher to allow him- or herself to become the object of critical analysis. In other terms, one must first analyze the transferences shaping the relations between the teacher and the students before one can work through personal and social transformation in the classroom.

Chapter 4

Life Is Beautiful, but for Whom?: Transference, Countertransference, and Student Responses to Teaching about the Holocaust

As I discussed in the previous chapters, a key to turning the popular depiction of the Holocaust and other cultural trauma into effective learning situations is to anticipate and counter the four main rhetorical modes of media reception often employed to block critical thinking and to promote counterproductive idealizations and identifications. To help clarify the presence of idealization, identification, assimilation, and universalization in the media and in educational efforts to analyze popular culture, in this chapter I relate these rhetorical modes of reception to how my students and writers on the Web have responded to the film *Life Is Beautiful*. One of my central goals in this analysis is to examine pedagogical methods instructors can employ to allow their students to work through their idealizing relationships with their teachers and their objects of study. Essential to this pedagogical work is an articulation of the psychoanalytic theories of idealization, transference, and countertransference.

LIFE IS BEAUTIFUL AND THE GLOBALIZING RHETORIC OF MEANINGLESS ENJOYMENT

One of the most interesting aspects of the popular responses to this film is that some people claim it is the best movie ever made about the Holocaust, while other people argue it has nothing to do with this historical trauma. What follows is a list of the most frequent responses (or variations on them) I have collected from my students:

> It is a comedy, so you are not supposed to take it seriously.
> I was very moved by it, and it had a profound effect on me.
> It was about a father's love for his son, and it had nothing to do with the war.
> It makes you appreciate the things that you have and how life is really beautiful.
> I didn't go to analyze it, I just relaxed and escaped.
> People like to see that even in a bad situation you can make fun of things.
> Everyone can relate to the love between a father and a son, so the film made the Holocaust more accessible.
> I don't know if it had a message, but it was really well done.
> It's really about the power of the imagination.
> It helps you to really experience the Holocaust.

Many of these responses center around the idea that this film is a comedy or a fictional form of entertainment, and thus one is not supposed to analyze it or take it too seriously. I categorize these responses as "universalizing" because they claim that the representation has no inherent meaning or value, and thus it is open to any and all interpretations. This rhetorical mode of reaction is the major form of resistance to critical analysis that one finds when one studies popular culture. Thus, even when a film or a television program is dealing with a historical event, people claim that we can learn nothing from it. In fact, many students argue that it is wrong to even try to analyze something that is merely there to entertain us.

This desire to see movies and television shows about the Holocaust as pure entertainment can serve the function of blocking any guilt or anxiety people may feel in relationship to this event. Thus, in his study of jokes, Freud argues that a major aspect of humor and comedy is to present serious issues and desires in a context where the joke teller and the audience will not take them seriously.[1] A comedy like *Life Is Beautiful* may therefore turn to history in order to negate the effect that history can have on us. Of course, one of the problems with this defensive rhetorical process is that people growing up in our current culture receive a lot of their historical knowledge from these fictional representations. Popular audiences thus can turn to historical reconstructions in order to assimilate history and in turn escape from their connection to the past.

Following the work of Slavoj Zizek, I have labeled this tendency to empty history and culture of critical analysis "universalizing" because it points to the definition of the modern democratic subject as a universal being without any content. According to Zizek in *Looking Awry*: "The subject of democracy is not a human person, 'man' in all the richness of his needs, interests, and beliefs. The subject of democracy, like the subject of psychoanalysis, is none other than the Cartesian subject in all its abstraction, the empty punctuality we reach after subtracting all its particular contents" (163).

Zizek argues here that the subject of modern democracy, philosophy, and psychoanalysis is empty and universal because the modern subject is supposed to be equal in front of the law regardless of race, creed, or gender. This definition of the democratic subject means that what we share as individuals in modern society is our formal equality, which ignores our personal beliefs and traditions.

In many ways, contemporary globalism can be seen as an expansion of this modern definition of the universal but empty subject. In fact, Zizek turns to a famous passage from Marx in order to link his conception of the empty subject to the globalizing tendencies of modern capitalism and science.

> Uninterrupted disturbance of all social conditions, everlasting uncertainty and agitation distinguish the bourgeois epoch from all earlier ones. All fixed, fast-frozen relations, with their train of ancient and venerable prejudices and opinions, are swept away, all new-formed ones become antiquated before they can ossify. All that is solid melts into air, all that is holy is profaned, and man is at last compelled to face with sober senses, his real conditions of life, and his relations with his kind (163).

Marx's description of the revolutionizing power of capitalism defines the modern subject as one who is constantly uncertain and changing, and therefore, this subject is void of any fixed internal or external attributes. Furthermore, since these subjects overcome all prejudices and beliefs, which traditionally block people from encountering the naked reality of economic exploitation, this continuously emptying subject is supposed to enter into direct confrontation with the realities of capitalism.

Zizek's turn to Marx here provides an important insight about how popular culture works in a globalizing society. Since capitalism is often the driving force behind cultural productions, and capitalism undermines all traditions, values, and beliefs, popular culture becomes a central source for both universalization and meaningless entertainment. In fact, my theory of the rhetoric of universalization tells us that behind Marx's and Zizek's definitions of modern capitalism, science, and philosophy rests the important developmental process of overcoming loss through symbolization and displacement. To be more precise, I posit that modern culture is built upon the human disposition to interact with the world through abstraction, symbolization, and repetition. A positive aspect of this universalizing process is that it creates order and predictability in our world; however, the price for generalized symbolization is often a loss of meaning, value, and belief. Therefore, one way of understanding Marx's idea that "all that is solid melts into air" is to link capitalist innovation to symbolic abstraction and universalization.

In terms of our contemporary globalizing world, what often unites us is a shared popular culture that we claim has no inherent value. In fact, the term *entertainment* is derived from the Latin word to connect and to unite, and as democratic citizens—all equal in front of the law—what we share in common is often based on a rhetorical lack of content or identity. Perhaps this is a key to both the positive and the negative aspects of our global community and economy. On the one hand, we are all supposed to be free to interpret things the way we want to and to be treated equally in legal and economic exchanges. On the other hand, this equality strips us of our differences and particular identities.

The globalization of American popular culture represents the promotion of a form of social interaction that has been rendered unconscious. Thus, it is not true to say that our cultural representations are void of meaning and value; rather, it is more accurate to posit that we internalize these messages in a nonconscious way. To emphasize the unconscious aspects of our globalizing culture to their students, educators can show them a supposedly meaningless film or television scene and then walk them through the various cognitive steps required to process the cultural information. The key here is to reveal to students how culture always promotes meanings and values. In order to escape responsibility and criticism, however, the producers of these representations attempt to claim that their productions have no meaning or value. The defensive rhetoric of meaningless entertainment is thus a universalized claim that is most often internalized by popular audiences on an unconscious level.

The Rhetoric of Identification in a Globalizing World

In order to counter this rhetorical claim of a lack of meaning in globalizing culture, many audiences turn to a second type of rhetorical resistance to critical thinking: identification. We find an instance of this type of resistance in the student response mentioned earlier that "Everyone can relate to the love between a father and a son, so the film made the Holocaust more accessible." In this type of response we see the way that people defend against history by concentrating on their own emotional responses to the historical representation. However, as I have shown in the previous chapters, just because people can empathetically relate to certain feelings, this does not mean they have learned anything about the causes and actualities of the events in question. In fact, this quick form of emotional release often relates to a false sense of dealing with the subject matter at hand.[2] For example, when students state that "It helps you to really experience the Holocaust" and "I was very moved by it, and it had a profound effect on me," we must ask what the real benefits of this type of vicarious experience are.

One way of problematizing this rhetoric of empathic identification is to have students compare their own subjective responses to Holocaust representations to historical texts concentrating on collective human suffering and collective resistance to oppression. By simply asking students to compose anonymous free writes that try to account for the differences between their own personal responses and the responses of groups to historical traumas, teachers can help start the process of relating individual rhetorical defenses to larger cultural and social issues. Similarly, by shifting the focus from the individual to the collective, specific writing assignments can push the idea that social change often happens through collective action, not just individual responses. In other words, teachers interested in promoting positive social change have to actively work to counter the contemporary movement to privatize and individualize culture and history.

In many of the identifying responses listed earlier in this chapter, the students return to themselves as the source of identity and emotional release.[3] The underlying idea driving this privatizing rhetoric is that it is through the experience of the "Other" that people gain access to their own true feelings. Yet, one of the problems with this form of empathy is that it tends to deny the reality and the history of the person and events being portrayed in favor of the emotional responses of the viewer. Thus, many of my students used the rhetoric of empathic identification to argue that the film made them value their own life more and made them realize it is always important to stay optimistic.

Psychoanalysis tells us that these types of idealizing and identifying responses work to create an unconscious transference between the object and the subject of study. According to analytic theory and practice, this mode of relationship is a fundamental and necessary way of interacting with the world, but left unanalyzed, the idealizing transference is a mode of resistance that blocks critical analysis and personal responsibility. From a pedagogical perspective, transference means that types of resistance are both the problem and the key to many solutions. Note, however, that most modes of pedagogy simply repress the importance of such resistance and act as if it were not present. Yet, any teacher who is really concerned with student learning cannot simply ignore the most vital driving force behind the educational relationship. Still, since the majority of teachers are not trained analysts and most educational situations are not analysis, an effective use of psychoanalytic ideas and practices in non-psychoanalytic relationships has to be developed.

As I have been advocating throughout this book, the move from the discourse of psychoanalysis to the discourse of rhetoric can help teachers educate their students about psychological defense mechanisms without having the students feel that their identities are being attacked. In this context, rhetoric offers a general accounting of social and personal resistances, and I have found that students are much more willing to analyze their own rhetoric than their own defenses. A turn to rhetoric also helps keep the dialectic between individual psychology and social ideology open and alive. The importance of this dialectic can be seen in the way that so many students will resist learning new information about themselves or their society if they feel they are being told that their subjectivity is being socially determined. Instead of accepting the idea that their identities are being shaped by larger forces outside of their control, many students will respond by either shutting down or returning to their selves as the center of their causality. Much of popular culture reinforces this re-centering on the self. Moreover, a psychoanalytic understanding of subjectivity tells us that the self is an ideal and imaginary form built out of idealizations and identifications. Jacques Lacan goes as far as arguing with that the ego is essentially a symptom or defense mechanism. It thus makes psychological sense for students to resist theories or educators who tell them their identities are socially determined or constructed.

Students therefore need to be given a space in which to express their identities and a place where those identities can be questioned, and the rhetorical analysis of defense mechanisms in anonymous writing assignments often provides an important pedagogical location for making social theories more effective. I also recommend here the analysis of popular reactions to popular culture as a way for teachers to model for students the analysis of rhetorical resistances. Equally essential, however, is constantly bouncing between the rhetorical analysis of social texts and the critical analysis of individual responses and resistance.

For example, I often bring into class a list of past student responses to the cultural productions I analyze during the course. I then discuss with students the different ways of categorizing these responses through their shared rhetorical moves. For example, in order to explain the rhetoric of idealization and identification, I point out all the responses that seem to return to the audience as the object of the analysis. This shows how, in the rhetorical act of refocusing the effect of the film onto the effect it has on oneself, a viewer often ends up idealizing both the creator of the film and his or her own personal values. In this way, a film about historical horror and tragedy can be viewed as a personal story about the greatness of human courage and love. While I do not deny that these aspects were presented in *Life Is Beautiful*, I do think it is important to show students why it is reductive to center one's reaction solely on one's own emotional and personal response. In fact, for many viewers this film acted as a mirror in which they saw their own ideal selves reflected back to them.

This type of ideal narcissistic reaction often hides the particular experiences of others behind a false wall of universal suffering or celebration. For instance, when my students posit that anyone can relate to this story of love and pain, they often imply that the specific historical facts and personal situations are not important. In these acts of rhetorical universalization coupled with empathic idealization and identification, the importance of the Holocaust and its differences from other historical events are denied. Yet, since mass culture must make itself "popular" for a wide range of people and values, the best way to do this is to deny difference and specificity and try to the tell a story that "anyone" can relate to. Furthermore, in order to appeal to the idealizing tendencies of the universal audience, popular movies often rely on showing off their technical skill and special effects. In this rhetorical move of focusing the audience's attention on the aesthetic production of the film, the actual information and the messages that are being communicated are often obscured. Thus, when my students claim "It's really about the power of imagination," what they are saying is that the idealization of the artist is more important then the content of the film. This is not to say that people do not learn from popular culture; on the contrary, they learn a lot. The problem is that people are not always aware of what they are taking in when they watch a film for pure enjoyment or a sense of escape.

In order to teach about the Holocaust in this type of cultural context, one must actively work against the message that popular culture has no meaning and that art is only for art's sake. For, to say that a film that is centered on fascism and life in a concentration camp has no meaning is to deny the value and importance of the victims and survivors of the Holocaust. Furthermore, I would argue that this mode of postmodern Holocaust denial is neither intentional nor direct, but it still has some of the same effects of the more blatant modes of Holocaust denial. For example, this movie was able to claim the power and prestige of a film about an important historical period without being historical at all. In fact, some people who admire the film often proclaim this lack of a historical context. This "decontextualizing" aspect of the movie is captured by the student response "It was about a father's love for his son, and it had nothing to do with the war." This denial of the historical specificity of the film relates to the general way that our culture rips events and symbols out of their original contexts and places them in new contexts.

I call this rhetorical mode of resistance "assimilating" in order to stress the fictional and adaptive nature of many cultural representations. To assimilate a symbol or event is not only to absorb something into a new context but also to play upon recognizable themes and attributes for a generalized audience. Likewise, the person who tries to assimilate is someone who attempts to copy and mimic the dominant characteristics and beliefs of a given culture. Moreover, in order to take on these borrowed attributes, the assimilator has to first remove these signs and symbols from their original context and redeploy them in a new context.

Analyzing Web Responses

Teachers can use this model of assimilation in order to discuss the dominant modes of representation in our current culture and educational systems. Importantly, I have found that the power of peer culture and popular culture to shape students' defense mechanisms makes it necessary to examine in class both the students' responses to poplar culture and the ways popular culture calls for certain responses. The pedagogical model I articulate here moves between analyzing how popular audiences actually respond to a particular cultural object and how that object caters to a particular rhetorical defense mechanism. In turn, my notion of assimilation studies the ways people assimilate to pop cultural representations as well as how pop culture relies on assimilating preexisting stereotypes, generalizations, and prejudices.[4] The paradox of assimilation is that it involves people copying shared cultural representations at the same time as the culture is assimilating past generalizations.

One of the best ways to examine this dialectic of assimilation with students is by analyzing the discussion of culture on the World Wide Web. For example, in the Yahoo discussion group about *Life Is Beautiful*, one can find a wide range of popular responses presented in an anonymous and noneducational environment.[5] I often have my students research this site to collect examples of popular rhetorical defense mechanisms. For instance, one student brought in this response: "I know it was not realistic . . . The story showed that man is also capable of purity and goodness, even in the face of evil. We do not know if they were in the camp for a week or a year, where the camp was located, or any of the facts . . . we draw our own conclusions. We do know that pure love exists."

My student noted that this writer starts by referring to the fictional nature of the film. Thus, according to this rhetorical move, it is not a realistic movie, so we should not think about it in a realistic fashion. In fact, its lack of realism and specificity allows the film to take on a universal appeal. In turn, the universality of the unrealistic picture produces an idealization of love, purity, and goodness in the form of the ideal relationship between the father and the son. Finally, this idealizing process produces an empathic celebration of pure love. The movement of this viewer's logic therefore passes through the four major rhetorical defenses I have been discussing. Moreover, this combination of idealization, identification, assimilation, and universalization transfers the focus of the film and history to the emotional experience of the individual viewer. Of course, one of the major problems with this approach is that it often leaves history, critical thinking, and the suffering of others by the roadside.

Walking my students through the different rhetorical turns in the previous statement enables me to model for them a way of locating and analyzing the dominant rhetorical moves of interpretation and resistance. Once again, my pedagogical method here is to first use the relatively safe space of other people's rhetorical defense mechanisms in order to help students gain a vocabulary and conceptual framework for understanding their own defense mechanisms and reading strategies. Furthermore, one of the first conceptual steps in this pedagogical process is to get students to recognize the serious nature of seemingly unserious popular culture. In fact, this method follows Freud's important insight that the supposedly lowest aspects of human culture—dreams, jokes, fantasies, and mistakes—provide important information for personal and cultural understanding. [6] Thus, by combining high theory with low culture, we are able to provide a more integrated cultural studies and rhetorical pedagogy.

While many students and Web discussants cling to the idea that popular culture is inherently meaningless, many also claim that the power of culture to produce empathy and identification makes it highly meaningful and personal. These viewers argue that instead of a quick move to universalize the subject matter and forget about the Holocaust, this form of popular culture allows one to combine important knowledge with personal investment. For example, one response claims, "The way the movie was directed made you feel as if you were in the camps along with the characters." In this form of empathic identification, we witness a merging between the audience and the sufferers of history through a process of emotionally reexperiencing the pain of others. Moreover, one identifies with the "characters" in the camps and not necessarily with the victims of the Holocaust. In other words, the viewer feels that he or she is really part of the constructed fiction and not the events of history.

However, students need to understand that this empathy with fictional characters may still have the negative effect of blocking critical thinking and analysis. Therefore, although postmodern culture asserts that to reexperience something is the same thing as actively learning about it, it is clear from the responses posted on the Internet, and the responses of my students, that on a conscious level they often have learned very little about the actual events being depicted in this movie and other popular culture productions. Furthermore, this use of history in order to not learn about history is evident in many of the comments my students have collected. For example, one writer declares: "I think . . . I responded to the emotional landscapes that the director created in 'Life Is Beautiful' as opposed to other films in which the holocaust is portrayed with the focus on the violence and the horror/torture which I tend to shut off from and generally become numb to those types of images." Whereas this writer describes the way that violence in film can undermine the ability of someone to react to the content of the movie, one has to question what can be learned from a picture that denies the true violence of the history that is being depicted.

As one astute writer in this online discussion group points out, this act of presenting a violent period of history without much violence is similar to the ways that the father Guido in the movie tries to protect his son from the horrors of the Holocaust. This discussant posits that the viewer is placed in the same position as the young boy who must be shielded from the true horrors of the camp. She also argues that this way of pacifying the audience is "condescending" and "It assumes that a parent—first and foremost—has the power to control and present situations in the way they want to for their children. The whole point about such horrors like the Holocaust is that you can't protect your children." I feel that this writer does engage in an empathic relationship with the film and the sufferers of the Holocaust, but she is also able to maintain a level of critical distance and see that the parents at the camps could not protect their children. In fact, she does universalize this point by stating that no one can control what his or her children are going to see, but this universalization occurs after she recognizes the differences of the historical contexts.

Many people react with outrage and dismissal in response to this writer's critical comments. One strong proponent of the film declares: "YA AIN'T UNDERSTOOD NOTHIN' OF THE MOVIE! This movie is NOT a movie about the holocaust! It wasn't supposed to show exactly what happened in the camps! Schindler's List did that already! This was supposed not to tell your children that the holocaust was bad, but that LIFE is beautiful, and to always enjoy it!" These statements are based on the idea that there can be only one reading of a film and that reading is determined by the intentions of the idealized director. Furthermore, this writer claims that *Schindler's List* already covered the Holocaust, and thus there is really nothing else to add.

Anyone who has ever taught literature knows that this concentration on the one intention of the author is extremely hard to shake. The idealization of the one meaning of the author usually entails an idealization of the person who understands this singular meaning. Many times this type of argument relies on the idea that history is made by great individuals, not by cultures and group actions. Importantly, this *I*dealizing transferential perspective insists that we should never criticize a work of art; rather, we are supposed to celebrate its greatness and our greatness—with which it puts us in touch. One of the problems with this mode of idealization is that it not only misreads history but also tends to attack anyone who voices a critical perspective. Thus, one discussant argues: "You missed the whole sense of the film as it was the intention of Benigni and whoever I heard that saw the film has come to appreciate." This viewer believes there can be only one meaning for the film, and that one meaning is based on the intentions of both the author and the audience. Anyone with an alternative view is just blowing hot air. I have found that by highlighting some of these anti-analytic statements that appear on Web sites, students are better able to work through their own resistance to critical analysis.

Lacking History on the Web

In one of my writing and research courses I had my students analyze the debates and commentaries that were distributed on another Internet site called epinions.com.[7] After reading and writing about the Holocaust for several weeks, my students were shocked by the lack of sensitivity and knowledge shown by the general public regarding anti-Semitism and the Holocaust. What bothered my students the most was the complete lack of historical specificity and information these comments displayed. For example, the following "epinion" replicates many of the subtle historical distortions circulating on the Web:

> Then the war starts, and thus, so does the persecution of Jews. He unfortunately, is one of those taken in, along with his son. In the concentration camp, the kid is hidden away in the overfilled dormitory, so that his father can make sure of his safety. Throughout, he must keep his son's spirits up, and lift him out of his boredom and depression. Thus, this film becomes not just a comedic tale, but a serious drama composed of family bonds and keeping your spirit against all odds.
>
> One of the most commendable aspects of this film is that it was not another war story that was depressing, methodic or violent, but a story that focuses on characters and personalities, irregardless of whether they are placed in a war situation or not. Though we see Benigni's depiction of the father go through the most difficult of situations, we see him pull through, and fight back without any bitterness but with invention, intelligence, and comedy. In short, this film focuses on the human character, the human will, and a father's love.
>
> "Life is Beautiful" will make you think exactly what life is really all about, and hopefully by the end, you will realize that life is about protecting and nurturing those closest to you, to the end. I loved this movie, and I hope that you will, too.

This analysis of the film stresses the development of the characters and the important need to protect one's family. However, as my students pointed out, wasn't the Holocaust a situation where families were separated and one could do nothing to protect them? Furthermore, how could one develop one's sense of self in a situation that dehumanized people and stripped them of all sense of dignity and control?

Since this movie (and, by extension, the commentator as well) effaces all historical realities and conditions from the Holocaust, it becomes easy for the viewer to project his or her own values and desires onto the leading characters. In this sense, it is reasonable for the commentator to praise the father's relationship with his son, but one has to wonder why he also lauds the father's lack of bitterness. In a very subtle way, these comments seem to blame the victims of the Holocaust for being bitter and not as inventive as the father in the film. In fact, many people who lack knowledge concerning this time period wonder why the Jews did not resist their murder and imprisonment. Since this film, and others like it, ask the viewer to empathetically identify with the main characters in a universal story of redemption, it is not unreasonable for the viewers to apply the situation of their own lives to the situation of this totally different context. Furthermore, *Life Is Beautiful* calls for this simplistic mode of identification by decontextualizing the setting of its own narration.

It is interesting to note that when people did present a negative view of this film on the Web, they were quickly attacked for not having a sense of humor or for being a negative person. Thus, just as the film tries to put a happy face on an unhappy situation, film viewers often demand that their fellow viewers stay positive. This idea is evident in the following opinion:

> Poor Roberto Benigni. There has been a terrible backlash against this brilliant actor/director. People said they got sick of seeing him bounce around and kiss everybody. I happen to find him refreshing and funny. Sure, maybe he overdoes it for the cameras, but people must be cynical to hate a person for being "too happy." In much the same way, people who don't appreciate Life Is Beautiful for the stunning, heartbreaking film that it is must also be horribly cynical or they just haven't seen the film. Maybe they haven't seen it because they think it is wrong to make light of the Holocaust. But if there was ever a film that takes the Holocaust seriously it is this one. I came away from this film with a greater appreciation of the effects of that war than I had with any other movie. Yes, there was laughter, but the audience knows the laughter that Benigni brings to his child is in order to mask the horror. Nobody in their right mind could possibly think this movie was just about the jokes and "Ha, ha, this concentration camp is so fun, let's all laugh." Anybody who thinks that has to be on crack. When I saw Benigni's face go from playful laughter while talking to his son, to terror and worry as he thought about what was

really happening, I realized that I was watching one of the greatest films ever made.

The message of the film, that it's important to remember that life is beautiful even under the worst of circumstances, is perhaps too positive a message for some people to handle. We're so used to wallowing in our misery that we have forgotten how to find happiness. The critics of this film's humorous aspect would do well to remember not to take themselves so seriously and that humor is often a healing agent.

At one moment this writer claims that the film is not a comedy, and then she goes on to attack people for not being able to take a joke. One reason for this possible contradiction is that she wants to defend this film against her own unconscious awareness that it may not be a good idea to turn the Holocaust into a comedy. Moreover, her desire to criticize anyone who criticizes the film points to a central aspect of our current culture: We often see that the person who points to a particular social or cultural problem becomes labeled as the problem. For instance, feminists are called sexists for pointing out gender disparities, and critics of racism are called racists for showing how race still functions in America. In the case of the Holocaust, people who criticize a Holocaust film or museum are called anti-Semitic.

Of course, one of the biggest problems with the film and the previous comment is the notion that the Holocaust should be used as a situation to prove that life is great. Perhaps as a defense against despair, people want to see good and happiness in a place where it does not belong or was not even possible. Or one could argue that people are tired of different ethnic groups complaining about their prior bad treatment, and so the dominant group wants the dominated groups to just be positive and forget the negative things that have happened in the past. While I hope that students themselves will make this connection between psychological resistances and larger cultural and political influences, I have found that one of my main tasks in this pedagogical process is to ask questions that seek to connect the psychological to the political, the cultural, and the rhetorical.

I also present examples of popular responses that provide positive models for critical analysis. For instance, in the following response, we see how a person's prior knowledge of the Holocaust affects the way he views popular culture representations of this event.

> I fear that this movie was critic-proof for several reasons, not the least of which being that writing a bad review of it will most likely be seen as approving of the Holocaust in some fashion. Let me just say right now that I am in no way disparaging the concentration camps or what happened there - I believe it, I am unable to comprehend it, and that's precisely why this movie left such a bad taste in my mouth. It is a dangerous, dangerous thing to make an allegory out of the Holocaust, and yet Benigni thought it would be in some way moving to de-emphasize the horrific details of the camps and focus instead on one man's devotion to his son. Make no mistake, this is Holocaust Lite—read Primo Levi's "Survival in Auschwitz" to get an inkling of what it was *really* like. I'll say it again—I cannot comprehend the inhumanity of what took place during the war, and no one who wasn't there can either. So why is it OK for Benigni to make a movie that depicts the camps as a place where children could survive and piles of bodies were only dimly visible in the mist? The tremendous outpouring of praise that this movie got was a disturbing indication of how much people want to believe these scenes. I think that by serving up Holocaust Lite, Benigni actually made this monstrous, inconceivable thing accessible to the masses, suddenly palatable, like watered-down hot sauce. What most of us were unable to face was presented to us in a neat, digestible package, complete with humor and a "happy" ending to make it go down smoothly. Shouldn't this be frightening? Shouldn't we all walk away from a movie about the Holocaust sick and exhausted with grief? Should it be permissible to combine such elements this way? This is one of the greatest crimes against humanity in the history of the world—genocide does not deserve to be sugar-coated.

This comment begins by arguing that one can be critical of this film and still not be a Holocaust denier. It is clear that this defensive position is derived from the fear that one should not go against popular opinion. Yet, this commentator does brave the storm, and we can see how his argument derives in part from his reading of Primo Levi's autobiographical text. In other words, unlike many of the other discussants, he does not compare this film to other pop culture Holocaust representations; rather, he relates *Life Is Beautiful* to a novel that is full of real details and historical contexts. While I am not arguing here that film is by definition inferior to books, what this commentary does show is how a book can provide for both a more accurate description of a historical context and still leave space for the limits of representation.[8]

In response to this man's criticism of the film, many negative comments were posted. One of the people who disagreed with him also turned to the field of literature to make her argument.

> Guido knew and we knew the horrors of the camp but he was determined that his son would not know them. I've read many, many books on the Holocaust and most of the victims tried to believe in the basic goodness of man (Anne Frank is the most quoted example). I don't feel this was trivialized or made light of. In fact the contrast between what Guido told his son and the truth was painfully obvious to all but Joshua, a four or five year old boy. It is that contrast that shows the horrors. I'm sorry to disagree with you on this so strongly. You have a definite right to your opinion and I hope you will consider my comments. Thanks! I'm sorry to disagree with you so totally, Mr. Alexander, but I couldn't pass your review and not say something.

This comment is in part inspired by *The Diary of Ann Frank*, which I would argue does present a more positive image of human beings than most novelistic depictions of the Holocaust. For, I think, it would be untrue to say that most novels concerning the Holocaust depict the idealized Christian message of the essential goodness of all human beings.

In another critical response to the discussant's negative criticism of this film, we find a different usage of literature to defend the movie: "There are many books and movies on the ins and outs of concentration camps and the Holocaust, so we pretty much know how bad the Holocaust was. This was supposed to be a different take on it. And in response to your argument that critics were afraid to say anything negative about it, I refer you to 'Jakob the Liar.' Just wanted to tell you, and I also respect your opinion. I like people who break away from the pack. Keep up the good opinions."

Here we are told that everyone knows enough about the Holocaust already, and so we do not need any more films or books on this subject. And like the previous writer, this discussant honors the right of everyone to his or her free opinion. In fact, this writer praises the critic for not conforming to the dominant view of this film. Once again we find in these statements a curious blend of the celebration of individualism, the universal right to free speech, and a desire to stop learning about history.

In my turn to the Internet as an archive of popular opinions, I hope to show my students how the World Wide Web is founded on the universalizing rhetorical principles of free speech, individualism, and decontextualized histories. As a result, however, it is often hard to find a thoughtful discussion of the Holocaust on the Internet because the Holocaust itself represents a historical situation where individualism and free speech were, most often, either curtailed or destroyed. Still, I do not want to give students the impression that I consider all discussion on the Web to be useless and that it is impossible to represent the Holocaust in film or in discussion groups. In fact, there are many informative sites on the Web.[9] Likewise, for our purposes here, the Web provides vital insights into how a popular audience is shaped by the rhetoric of idealization, identification, assimilation, and universalization. By addressing these defense mechanisms, we can help our students and others work through the media and political manipulations of trauma and history.

Educational environments are themselves structured by the same rhetorical defense mechanisms that shape popular culture, however, and therefore any critical pedagogy must also work to examine the institutional, psychological, and social factors that shape student learning. In order to help us work though these shared resistances in popular culture and contemporary education, I want to introduce the psychoanalytic concept of transference, which is a central way of understanding how the rhetoric of idealization functions. I also want to examine the concept of countertransference to posit pedagogical strategies for countering the various rhetorical defense mechanisms articulated here.

Transference and Idealization in Education

It should not surprise any teacher that students often place their instructors in the position of being the one who is supposed to know the subject matter. According to Lacan, this idealization of the teacher can help the student feel ideal and can also represent a type of surrendering to the will of the other.[10] In fact, one of Lacan's major contributions to the field of psychoanalysis is his idea that transference is essentially about knowledge, not displaced sexual desire. For Lacan, we love and idealize the person we presume to have knowledge, and this type of transference acts as a major mode of rhetorical resistance in and out of the analytic setting.[11]

In the context of educational institutions, the idealization of the teacher's knowledge can work as both a motivator for student learning and a major mode of resistance. Like the psychoanalytic relationship between the analyst and the patient, teaching can start only if the students place a certain level of faith in the knowledge of their educators. This process of transferring knowledge onto the teacher, however, can quickly result in the formation of a dependent relationship in which the teacher has all the power and knowledge, and the student simply sits back and watches the idealized instructor. To make education more effective and interactive, it is therefore necessary to disrupt the idealization of the teacher. One of the best ways to transform the student-teacher relationship is to have students read *and critique* the work of their instructors, and the key to this process is to create a situation and an environment where students feel safe analyzing their transferential relationships with teachers and objects of study.

One method I use to start this process of de-idealization is to have students engage in anonymous online conversations concerning my own writings about popular responses to cultural productions. For instance, I had students read an earlier version of this chapter and discuss online in an open and unstructured environment their objective and subjective reactions to my analysis of students' resistance. This process enabled students to challenge the authority of the teacher and to examine their own subjective responses to my educational efforts in the context of a safe space.

Some students wrote that as a professor, I do not know how to laugh at things and that I should lighten up and relax. I think this type of comment points to a generational divide between students, who are often used to seeing everything as a source for entertainment, and teachers, who still believe a strict distinction exists between serious subject matters and nonserious entertainment. In reaction to this kind of response, I think it is best to acknowledge it and not to try to correct it. In other words, teachers need to recognize the different social and cultural views of students. In fact, the way the teacher responds to student criticisms often determines the ability of students to really take a critical position on their own educational attitudes.

Another comment I have received which directly addresses the teacher-student relationship is, "The teacher seems to be attacking students and making fun of them." Many students commonly feel this way when teachers begin to question their pupils' ideas and attitudes. Once again, the teacher must recognize these feelings and not try to counter them. In fact, in my experience, when I do acknowledge this type of criticism, students begin to see me more as a human being and less as a teaching machine. In other words, by allowing the negative transference to surface, the subjectivity of the teacher and the subjectivity of the students are able to communicate.[12]

I am not arguing here that the teacher should be a verbal punching bag for the students. What I do want to stress, however, is the need for the teacher to block the ego-to-ego relationship, where the demonization of the other is matched by the self-righteous idealization of one's own opinions. By allowing for the expression of negative criticisms, the teacher can be de-idealized, and the transference becomes exposed. Furthermore, one of the goals of having students share their personal responses to the teacher in a safe and anonymous environment is that students are given a chance to recognize emotional aspects of the learning process. In fact, most educational settings function by discarding affective responses, and this pedagogical process of repressing affect can function to exclude important parts of student subjectivity and academic research.[13]

Psychoanalytic theory and practice also inform us that another way of transforming the teacher-student transference is through the use of countertransference. I use this term to invoke the various strategies a teacher or student can use in order to provide an objective account of their own subjectivity in relation to an object of knowledge, culture, and research. The psychoanalytic notion of countertransference explores a new model for scientific objectivity that moves beyond the "modern" educational stress on neutrality, abstraction, generalization, and impersonality. In this form of scientific discovery and learning, the creation of a "self-reflexive classroom" helps students to explore and utilize their personal unconscious negative reactions to their research material. Moreover, by coming to terms with their own subjective responses to various threatening subjects, students can learn to integrate emotional aspects of selfhood into the learning process.

In developing this notion of countertransference, we see how the scientific quest for an "objective" stance often forces students *and* teachers to repress and deny their unconscious fears and doubts related to the subject matter in question. This modern rhetorical repression of subjectivity often denies the researcher, teacher, and learner access to important information essential to the inquiry process. The student's and the teacher's desires, fears, and values are most often related to the particular object of study, and thus a denial of their own feelings only hides subjective reactions, which may unconsciously hinder or possibly help the learning process. For example, in my own research on social science explanations of the Holocaust, I have found that my strong emotional responses to this subject matter have helped block my ability to objectively analyze certain theories and concepts. On the other hand, I have also found that when I analyze my subjective responses, I am often able to use these emotional responses as indicators of unexplored areas of research.

This effort to incorporate countertransferential aspects of subjectivity into the writing of social science research has been explored by Alain Giami in his article "Counter-Transference in Social Research: Beyond George Devereux." He writes:

> A researcher's counter-transference can be defined as the sum of unconscious and emotional reactions, including anxiety, affecting his/her relation with the observed subject and situation. These reactions produce distortions in the process of knowledge construction that remain hidden from the researcher. Notions of "inappropriateness" and "resistance," as defined by Schimek, become central in understanding the cognitive processes affecting the researcher, because they highlight the researcher's reactions to aspects of reality emerging in fieldwork. Counter-transference points to the researcher's difficulty in clearly distinguishing material that comes from outside (the subject, the field) and from inside (his/her own emotional reactions). The researcher has to struggle with these emotional reactions and anxieties (online).

Giami effectively shows here why we need to take into account this notion of countertransference when we ask students to perform research or cultural analysis. The student, as researcher, is not the object of study; rather, the student researcher is a subject who must sift through a variety of internal and external information.

Drawing on the work of George Devereux, Giami posits that social science research always implies a dialectical communication between subjective and objective factors:

The researcher is, in one way or another, the subject and object of the knowledge that he/she elaborates. The specific position he/she occupies in the field allows at the same time for a specific kind of focus and for specific blind spots. From any one position, there are aspects of the world that one can perceive and aspects that one cannot. Absolute objectivity is, by definition, impossible and one has to find the appropriate focus, the "good distance," according to one's research object. The position of the researcher in the field defines (1) what he/she can know, (2) what he/she might be able to know, (3) what he/she cannot know and last but not least (4) what he/she actively refuses to know for some social or psychological reason. In some cases, researchers know what they do not know and what they cannot know; in others they do not even take account of what they cannot know (online).

By getting our students to examine their own and our own blind spots and limitations in relation to a particular subject matter, we can help them understand the subjective aspects of objectivity as they learn to monitor their own countertransferential reactions to their objects of study.

To help clarify this role of countertransference in de-idealizing the teacher, the students, and the object of study, I want to examine a turning point I experienced during a course I taught on the theme of "Social Science Approaches to the Holocaust, Prejudice, and Cultural Assimilation." Throughout this course, students often had a difficult time discussing and writing about the contemporary depictions of the Holocaust and the role of prejudice in popular culture. In fact, until I motivated my students to explore their own personal reactions to the class topic, they often resisted reading the class assignments and engaging in substantive class discussions. The turning point came when I took the risk of openly discussing my own reaction to the students' negative responses to the class readings and assignments. In a moment of frustration, I compared the passivity of the students to the passivity of the people who let the Holocaust happen. I later analyzed my own response as an unchecked subjective projection.

One of the unexpected benefits of revealing my subjective reaction to the students' resistance was that the students stopped seeing me as a purely "objective" teacher; instead, they were able to witness the role played by my own subjectivity in my quest for an objective analysis. Later on, many students said that the class discussion of my "countertransferential" feelings was a key to making them feel safe about expressing their own personal feelings and reactions. Moreover, once students saw that I was also "human" and that I considered myself a part of their culture, they were more willing to engage in serious research and writing. While I believe many risks are inherent in this type of teaching, I nevertheless feel that the best way to teach about "objectivity" is by analyzing the role of subjectivity in all learning processes.

In many cases this need to explore the subjectivity of the objective researcher requires the teacher/researcher to analyze his or her own subjective reactions to the class's dynamics. In psychoanalytic terms, this analysis of the teacher's reactions to the students' resistance can be called "countertransferential" because it involves the analyst/teacher/researcher projecting his or her own unconscious fears and desires onto the students and the object of study. For example, I often reacted to students' passivity by unconsciously equating them with the passive bystanders in Nazi Germany. Until I openly made this connection to the class, I was unable to deal with their resistance in a more rational and objective fashion. Furthermore, if the only way to approach a state of scientific objectivity is to try to account for and control most subjective factors, the analysis of the unconscious subjectivity of the students *and* the teacher becomes essential.

To help rectify this situation of repressing and excluding emotional aspects of student subjectivity, I discuss in the next chapter various ways that teachers can use the rhetoric of assimilation in an effort to motivate self-reflective student essays. This mode of writing and analysis is dedicated to the idea that students can analyze and transform their social and personal views if they first work through their own defense mechanisms. Furthermore, I posit that this kind of self-reflective analysis needs to be modeled by teachers who employ psychoanalytic cultural criticism not only to analyze texts but also to transform educational relationships. I also use the next chapter to demonstrate how contemporary politics and culture affect the possibilities of our learning environments.

CHAPTER 5

FREUD GOES TO *SOUTH PARK*: TEACHING
AGAINST POSTMODERN PREJUDICES AND
EQUAL OPPORTUNITY HATRED

As I have argued throughout this work, one reason to teach about the Holocaust in contemporary college courses is that this historical event can push students to take a serious look at the possible ramifications of ethnic intolerance, personal passivity, submission to authority, technological dehumanization, and the absence of democracy. However, I have also posited that contemporary popular culture often turns to the Holocaust in order to represent the opposite message: to proclaim it good to be intolerant, to reward passivity, and to encourage submission to idealized authority. Moreover, I have traced the growing tendency to treat the Holocaust as a laughing matter and to turn this tragedy into a source for personal idealization. What I hope to accomplish in this chapter is a more detailed analysis of how our culture often promotes an unconscious mode of internalized racism under the banner of universal intolerance and political incorrectness. The key to this explanation will be an expanded use of the term *assimilation*. In fact, I will turn to this rhetorical defense mechanism not only to show how people end up internalizing the very stereotypes and prejudices that oppress them but also to examine how contemporary students are shaped by a globalized discourse of cultural assimilation. Finally, this notion of assimilation will be employed to discuss the use of self-reflective writing assignments in an effort to break out of self-destructive rhetorical modes.

Examining South Park in a Backlash Culture

When I teach about the movie *South Park*, I want my students to consider the effects of prejudice, trauma, and the rhetoric of resistance in the context of popular culture and in their own daily lives. Therefore, before we watch and analyze this film in class, I examine the now-popular rhetoric of "equal opportunity hatred." The basic idea behind this contemporary idea is that if you represent prejudices against *everyone*, you are not negatively affecting *anyone*. In other terms, by universalizing (one might even say globalizing) hatred and prejudice, people believe they are not responsible for anything they say. Moreover, this usage of universal hatred is usually invoked to counter a mythical order of repressive political correctness. I have found that even the most liberal students have accepted the idea that the academic world is controlled by political correctness and that popular culture offers freedom and escape by being politically incorrect.

Like many other television shows and movies, *South Park* gains a great deal of its popularity by proclaiming itself to be intolerant of all forms of tolerance. It also completes this rhetorical reversal by seeing itself as tolerant of intolerance: in other words, as my students often report, its humor is generated by "saying what you are not supposed to say." I will argue here that this rhetorical reversal, where one is taught to be intolerant of tolerance and tolerant of intolerance, is part of a larger social effort to challenge and reverse progressive efforts to fight stereotypes and prejudices in American culture. Furthermore, this rhetorical reversal can also be seen in the political and cultural process of undermining the popular support for the welfare state, while calling for tax breaks for the wealthy. In this upside-down rhetorical world, minorities are now seen as victimizers and abusers of the welfare system, while the wealthy majority is positioned to be the victim of excessive taxes and reversed racism.[1]

Knowingly or unknowingly, shows like *South Park* feed this rhetorical reversal that influences so many students and makes it even more difficult to teach about critical thinking and social change in higher education. One reason why humor plays such a major role in this attempt to reverse the cultural representation of minorities and dominant groups is that comedy itself often works by reversing values and social positions. Thus, in the classic structure of humor, a man acts like a woman or a pauper acts like a king. Moreover, much of popular comedy today is based on the use of ethnic stereotypes and prejudices that are allowed to be recycled because the victims of these negative depictions are the ones making these destructive self-representations. The contemporary structure of cultural assimilation and immigration also calls for the internalization of negative self-representations.

Assimilated Stereotypes in South Park

The assimilation and internalization of self-hatred in *South Park* is represented through the character of Kyle Broslovski, a boy who is attacked in a humorous way for being Jewish. Furthermore, the representation of his character and his family play off of the most basic stereotypes concerning Jewish people. Importantly, one of the two main writers of this show is Jewish, and thus we are forced to ask the question, Why do Jews in our culture often participate in the recirculation of anti-Semitic stereotypes?

This question of internalized anti-Semitism offers students one of the keys to understanding postmodern prejudices because most modes of discrimination pass through three major stages. At first, prejudices are brutally applied through real acts of dehumanization and enslavement. The next stage of prejudice often involves legal segregation and state-sponsored discrimination. Finally, in a third stage, the objects of prejudice internalize the stereotypes by which they have been victimized.[2] In this postmodern stage, negative and positive stereotypes provide a ground for self-recognition and identity. Moreover, the mass culture industry reinforces these negative self-representations by basing characters on the largest available cultural generalizations. Internalized anti-Semitism thus plays into this logic by centering Jewish identity on cultural assimilation, victimhood, and self-deprecating humor.

However, whenever I point out in class the various stereotypes and prejudices in the show, students are quick to argue that every group is equally attacked, and no one takes the attacks seriously. Here we see a strong example of the rhetoric of resistance: from a universalizing perspective, students claim that equal opportunity intolerance is not intolerant and that popular culture has no meaning or effect anyway. Students often feel that the teacher who points to prejudice in culture and society is really the one with the problem. In this version of "shooting the messenger," the critic motivating students to become aware of the destructive nature of prejudice is seen as a highly sensitive person who does not know how to take a joke and does not accept the universalized notions of tolerance and intolerance.

Once again, I think it is important for teachers to recognize this cultural reversal without directly challenging these deeply held social beliefs. Nevertheless, students *can* learn how to detect and move beyond the various rhetorical mechanisms that our culture uses to reinforce and recirculate prejudices and stereotypes. The trick is to analyze destructive rhetorical figures in a safe transitional place that is neither too personal nor too alien. Popular culture comedies like *South Park* can offer us this transitional space, but we need to provide students with the critical tools and concepts to help them stop the flow of information and examine the specific cultural elements used to construct humor and identity in our society.

One rhetorical strategy that popular culture employs is the use of extreme exaggeration both to circulate prejudices and to deny the import of these negative self-representations. For example, one of the main stereotypes in *South Park* is the relationship between the overly protective Jewish mother and the oppressed Jewish son. In fact, the main plot of the movie is that Kyle's politically correct aggressive mother wants to attack Canada because two Canadian filmmakers have made a foul-mouthed film that has affected her son and his friends. In this narrative, the stereotype of the overly protective Jewish mother is exaggerated to the point that she threatens to cause an international war. Furthermore, this extreme depiction of a Jewish mother is presented along with a series of decontextualized references to the Holocaust and World War II: there is a "Canadian Death Camp," children join "La Resistance," there are charges of "crimes against humanity," and various representations of concentration camps.[3]

What, in part, links the internalization of anti-Semitic stereotypes to the subtle and not-so-subtle reminders of the Holocaust is the strategy of taking serious issues and trying to empty them of their initial meaning and value. For instance, in one scene that takes place in Hell, we see a morphing between images of George Burns, Gandhi, Hitler, Saddam Hussein, and Satan. From the postmodern perspective of the filmmakers, it is possible that there is little difference between someone who played God in a movie (George Burns), someone who was a leader of his people (Gandhi), a former totalitarian leader (Hitler), a hated current despotic leader (Hussein), and the Christian representation of evil (Satan). Since these figures are images cut out of any historical and cultural context, they seem to have lost all inherent value or significance. In fact, these images have become secondary texts that can be assimilated into new contexts for our humor and entertainment. Furthermore, the fate of these decontextualized images parallels the fate of immigrating people who must assimilate to a new cultural context by shedding the value of their previous cultural and historical traditions and beliefs.[4]

We can find a strong incidence of this logic of assimilation and internalized anti-Semitism in the title of the movie—*South Park: Bigger, Longer, and Uncut*. The subtitle combines allusions to censorship, the uncircumcised penis, and a sense of phallic power. Moreover, since the Jewish mother becomes the main proponent of politically correct censorship in the movie, an interesting equation emerges involving the circumcision of the Jewish male, the dominance of the Jewish mother, and the censorship of free speech. Here, the Jewish ritual of circumcision is blamed on the mother and attached to a loss of freedom. In fact, what is idealized is the non-Jewish, uncircumcised exercise of phallic power and free speech.[5] The assimilation and internalization of anti-Semitism thus results in the idea that the Jewish tradition of circumcising men—cutting them off from their free and manly expression of aggression and sexuality—is the root of all evil in the world.[6]

Since the movie declares itself to be bigger, longer, and uncut, we can assume that true phallic power comes from the ability to deny circumcision and censorship by identifying with the idealized free and non-Jewish male member.[7] The movie thus subtly plays on the standard cultural opposition between the powerful Christian male and the feminized Jew. In turn, this opposition is projected onto a political fight where the Jewish mother becomes the source for feminizing and censoring the victimized Jewish male. Furthermore, this demonization of the Jewish mother and the feminized Jewish male is hidden behind the general rhetorical defensive strategy of mocking the idea that the media influences children in a negative way.

According to the film's logic, the problems with our children do not stem from the fact that they assimilate the obscene and prejudicial representations they see in the media. Rather, the problem is that the Jewish maternal superego wants to censor and castrate the naughty boys for expressing their true desires. It is important to stress here that students are often quick to buy this conservative rhetoric, namely, that the true cause of intolerance in our society is the politically correct people who are trying to fight intolerance.

One reason why students often equate political correctness with intolerance and prejudicial intolerance with tolerance is that they have grown up in a culture shaped by a successful conservative campaign to reverse our understandings of prejudice and tolerance. A key aspect of this rhetorical reversal is the universalized notion that there are no longer any prejudices in our society, and thus anyone who points out prejudices must be the cause of prejudice. In the case of teaching about prejudices, this reversed rhetoric cannot help but enter the room, and I believe it is necessary for teachers to deal with this conservative ideology in an indirect way. Since a direct attack on the conservative effort to reverse racism will usually serve only to reinforce students' political ideologies and investments, it is important to examine this rhetoric in a nonpolarizing way. Furthermore, the first step in this process of educating against the rhetorical reversal is to critique the notion of universal tolerance and then show examples of intolerance in popular culture. Teachers also need to change the popular view of political correctness by critiquing its excesses and affirming the positive aspects of treating others with respect.

While many students will claim that *South Park* is a liberal, anticonservative show, it is clear that for the writers of the film and television program, political correctness is the primary evil they fight against.[8] Like the incorrigible children they portray and like so many of our students, Matt Stone and Trey Parker feel strongly that censorship is worse than hate speech and that free speech is the ultimate good and should be celebrated.[9] This desire to endorse a universal message of free speech coupled with their idealization of the unrestrained individual is apparent in the many interviews they have given to defend their usage of bathroom humor and politically incorrect stereotypes. For instance, in response to the question of why he stresses the Jewishness of the character Kyle, Stone (whose mother is Jewish) proclaims: "It just creates more opportunity for comedy. It gives us more things to make fun of and we just think it's funny."[10] What Stone does not say in this statement is why Kyle's Jewish identity is funny and how this humor relates to Stone's own secular upbringing in Texas and Colorado.

Why Do We Laugh?

To help my students think about the culture of political incorrectness and internalized racism, I have found it necessary to get them to reflect on how humor and comedy function. One of the first ideas I posit in my classes is the notion that humor often derives from real feelings of pain and anxiety that are then turned into a "positive" experience by entertaining others. This theory is clarified in the sections of Freud's *Jokes and Their Relation to the Unconscious* where he posits that the essential goal of comedy is to make others laugh, and the way that we do this is by allowing the third-party audience to be the passive spectators of a veiled act of sexual aggression (116).[11] In fact, Freud argues that the true goal of a joke is to bond with a third party and to bribe this social other not to analyze or criticize one's humor (119). Thus jokes, while based on a social act of bonding, also serve the purpose of releasing repressed urges and desires.

In the classroom setting, this notion of using humor and prejudice to bond can be examined by having students discuss their own, and their friends', experiences with using humor to create in-groups and establish out-groups. One effective short writing assignment is to ask students to recall some of their earliest experiences when they saw humor being used to humiliate or make fun of someone else. After they write these short essays, I ask students to volunteer to share their experiences, and this process usually opens up a lively discussion concerning the possible negative effects of humor in our culture. This topic also functions to bring into class the role of emotion in the shaping of how we bond and form social groups. Once again, I want to highlight that in traditional college classroom assignments, affective responses are usually downplayed and repressed in order to establish a more intellectual and distant mode of analysis. Yet, as we see in the psychoanalytic theory of humor, affect plays a central role in social interaction, and thus any educational process that represses emotions works to hide important information and to negate parts of the students' own subjectivities.

While I want my students to see how diverse ethnic groups have been the objects of comedic humiliation, Freud's theory of humor is dominated by references to Jewish jokes.[12] One reason that Freud gives for this choice is the debatable idea that Jews are more self-critical than other people, and thus they make better comics (133). However, as many students insist, in our culture all types of social groups are the target of humor, and in most cases the comic is from the same group that is being attacked. Central to this structure of in-group comedy is the rhetorical notion that it is okay to attack someone in your own group, but it is not right for someone from another group to attack your group. In many cases, this logic of the in-group attacking itself functions to circulate prejudice under a safe cover. Furthermore, one of Freud's central ideas about humor is his notion that the ethnic group members perform an act of self-mockery for a neutral party, and therefore, even when a person of a group mocks his or her own group, that presentation of intolerance is performed for the social and cultural other.

It is essential to point out that in this structure of humor, the basic process of assimilation is the appeal of the minority to the dominant culture's definitions and values. Furthermore, Freud insists that the object, or second party, of the joke plays the role of the social censor who must be avoided (116). In fact, for Freud the primary example of humor is a dirty joke that is meant to seduce a woman, but due to her high moral standing, the joke must be redirected toward a male third party. In this structure, the third party becomes the ally of the first-party joke teller against the second-party female object and social censor. If we now apply this structure to the general framework of the film *South Park*, we see that the Jewish mother represents the object of the joke and the source of social censorship.[13] In order to overcome the Jewish mother's resistance to the pure expression of sexual aggression (the desire of the joke teller), the mother must be attacked for the benefit and the enjoyment of the third party, or audience. The joke teller and the audience thus bond over the attack on the Jewish mother.

Through this theory, students can begin to see how so much of our popular culture is often centered on a process of male bonding through the tragic or comedic stigmatization of minority groups. Even if the audience is not entirely made up of men, this theory argues that the viewer is placed in the third-party position of being the one who accepts or rejects the presentations of the first-party joke teller. Here, the third party represents the dominant culture that must be bribed by the first-party's victimization of the second party. In the case of *South Park*, we can say that Matt Stone (a Jew) victimizes his Jewish mother and identity in order to bond with his audience. Internalized anti-Semitism therefore serves the processes of assimilation by sacrificing a part of the subject's own ethnic identity in the goal of bonding with the dominant culture.

To help students understand this complicated three-party structure of humor and assimilation, it is important to relate this theory to their own personal experiences. However, the students often do not want to discuss this uncomfortable subject matter because they do not want to point to their own vulnerabilities and aggression. Instead of simply shying away from this difficult pedagogical problem, I propose using anonymous writing assignments or online discussions through which to express subject areas that are often repressed. In fact, a key aspect to Freud's work and psychoanalytic cultural studies is to dare to go to places that are uncomfortable to examine.

In contemporary culture, two of the most uncomfortable topics are homosexuality and homophobia. In fact, homophobia is often considered by popular culture to be the only prejudice that can be openly affirmed in public discourse. Furthermore, so much of our popular humor is driven by homophobic jokes, and this use of homophobia usually points to the desire of men to bond on an emotional level by stigmatizing homosexuals and affirming their own heterosexuality. In the context of the *South Park* film, the running homophobic joke is that Satan and Saddam Hussein are homosexual lovers in Hell. On one level, I would argue that this attack on homosexuality helps men to bond with each other through a process of sharing the rejection of same-sex desire through identification and demonization. However, if most humor is really about male bonding, then we must see the use of homophobia as a way of both revealing and hiding the true bonding that is going on in the film. For even if this film attempts to attack all forms of political correctness as a restraint on universal free speech, it defends this universal value from the position of bonding with the assumed universal male audience.

The Myth of Free Speech

In response to this theory of male bonding and homophobia, many students argue that we do not have a society controlled by men and that women are also comedians who make jokes to bond with each other. While I do not deny aspects of this argument, I point out to students that the rhetoric of universal equality can work to veil important inequalities in our society. In fact, students need to see how the hidden agenda of male bonding helps to explode the myth of the neutral realm of universal free speech and tolerance. Thus, as Stanley Fish argues in his book *There's No Such Thing as Free Speech*, the claim for a universal tolerance of all expression is always grounded on a hidden agenda of particular vested interests (7). In fact, Fish posits that all universal claims are invalid because they do not take into account the context and history of their own formulations (viii). Therefore, Fish posits that we must always contextualize every universal claim to see what interests lie behind it. In the case of *South Park*, the universal rhetoric of free speech relies upon the unstated idea that words have no real effect on people, and thus they should never be constrained. However, this idea is itself challenged by the notion that the words of the politically correct *do* actively constrain the freedoms of the politically incorrect. Yet, the way out of this conflict is to argue that unlike the words and actions of the proponents of political correctness, the politically incorrect makers of this movie do not believe their representations have any meaning or context.

For example, by making references to the Holocaust without any concern for the original context of these representations, the writers are able to claim that these depictions are not harmful to any particular group. Likewise, the representation of Kyle's Jewishness is seen as being purely entertaining and not anti-Semitic because the writers believe their representations have no value or effect. One reason they can make this claim is that they do not believe ethnic identity itself has any value or meaning other than its ability to make people stand out from the dominant crowd and be laughed at. Within the context of *South Park* and the ideologies affecting student subjectivities, this strategy of denying the value or import of ethnic identity and other cultural influences is a key to the idea that popular culture is really about nothing. In this sense, what Americans seem to value the most is the idea that our culture is meaningless, and our words and representations have no real effect. From this perspective, there is no such thing as hate speech, and what is really wrong about politically correct people and teachers in general is that they take words and representations too seriously. In a strange way, America is the most Zen-like culture around because what we value the most is our ability to spend a great deal of time experiencing nothingness and non-meaning.

IT'S ONLY A JOKE

This Zen-like philosophy within popular culture represents one of the strongest convictions of the audience.[14] Students often steadfastly cling to their right to meaningless entertainment, and they often interpret any attempt to contextualize or interpret popular culture as a horrible act threatening to rob them of their most cherished value. How do we then reconcile these two opposing claims? On one side, we have the argument that popular culture has no meaning or value, and on the other side, we find the argument that nothing has more value than the defense of popular culture and the freedom of expression.

Freud's theory of jokes helps us resolve this conflict by posing that the main function of jokes is to present serious issues in a manner that shields them from any type of criticism and analysis. Like free speech, humor thus creates a responsibility-free zone where people are given the opportunity to state anything they like without fear of being censored or restrained. Yet, humor itself needs restraint because it is generated out of the conflict between infantile desires (pure sex and aggression) and social norms. In order to hide this conflict, the first-person joke teller must rely on the third-party audience's ability to process the information of the joke while denying the value of the same information. For example, ethnic jokes rely on cultural stereotypes that the audience must recognize and understand but not acknowledge as being meaningful or valuable.

This Freudian theory of humor helps define the relationships among assimilation, popular culture, and internalized prejudices. In the context of assimilation, people first reduce their ethnic identity to the level of a stereotype, and then they tell the dominant culture that this representation means nothing. Thus, Matt Stone says that people like to laugh at the stereotypical representations of Jewish characters, but at the same time he claims that these representations are only entertainment, and they have no real meaning. Freud is able to account for this contradiction between the value and the meaninglessness of stereotypes through his notion of preconscious representations. This theory is one of the most misunderstood and neglected aspects of his work because it breaks down the simple opposition between unconscious and conscious ideas. By saying that stereotypes are preconscious, Freud indicates that we use them without our conscious awareness of them, and so they take on an automatic quality as if they were coming from some foreign place. Therefore, we feel that we are not responsible for these preconscious representations because we are barely aware of them, and we do not use them intentionally. Yet, I would argue that in popular culture, these types of preconscious prejudices are the most prevalent since people assimilate and circulate ideas they claim are not their own and have no meaning. Furthermore, it is ridiculous for people to say they have no prejudices, since our culture itself is predicated on the ability of a mass audience to recognize generalized traits of characters and ethnic groups. What we often assimilate in our culture are serious preconscious representations and generalizations that are placed in an unserious responsibility-free zone.

The Rhetoric of Denial

One reason why a film like *South Park* is such an effective pedagogical tool is that it helps reveal the popular rhetorical methods employed both for assimilating stereotypes and for denying their value and responsibility. A way that the writers accomplish this task of removing themselves from any responsibility for their representation of prejudices is to make the scapegoating process itself a ridiculous aspect of the film. Thus, *South Park* includes the song "Mountain Town" that openly anticipates the critics of the movie, and the movie within the movie:

> Off to the movies we shall go
> Where we learn everything that we know
> Cause the movies teach us what our parents don't have time to say
> And this movie's gonna make our lives complete. . . .

While these lyrics do indicate that children are highly influenced by the media, the words present this argument in such a stupid-sounding song that people are signaled not to take this idea seriously. In fact, this blaming of the media is then transformed into the humorous technique of using Canada as a scapegoat in the song "Blame Canada":

> Blame Canada! Shame on Canada!
> For the smut we must cut
> The trash we must dash
> The laughter and fun must all be undone
> We must blame them and cause a fuss
> Before somebody thinks of blaming us!

Like the song about the influence of the media, this song posits that parents blame other people in order to avoid their own responsibilities. The film does send important messages, but it then undercuts these same messages by placing them in contexts where the audience is told not to take them seriously. Therefore, by making fun of the way that people blame the media and other people for their own problems, the movie is able to remove all responsibility from its own representations.

Within the film's context of scapegoating and the denying of responsibility, we find several references to the Holocaust. For example, as the war between the United States and Canada heats up, a news anchor delivers the following address:

> A full-scale attack has been launched on Toronto, after the Canadians' last bombing, which took a horrible toll on the Arquette family. For security measures, our great American government is rounding up all citizens that have any Canadian blood, and putting them into camps. All Canadian-American citizens are to report to one of these Death Camps right away. Did I say "Death Camps"? I meant "Happy Camps," where you will eat the finest meals, have access to fabulous doctors, and be able to exercise regularly. Meanwhile, the war criminals, Terrance and Phillip, are prepped for their execution. Their execution will take place during a fabulous USO show, with special guest celebrities.

The mention here of "Death Camps," "fabulous doctors," and the execution of "war criminals" is intended to reference and mock different aspects of the Holocaust. Once again we must ask, Why would these writers do this? Are these references made because they are so recognizable, or does the humor set out to make light of the true traumatic nature of the Holocaust?

One possible way of responding to this question is to look at the way the Holocaust is tied to Jewish identity in our culture and the way this identity is both affirmed and denied in the film. In the case of Kyle, one of the running jokes on the show, and in the movie, is that it is not his fault for being Jewish. We find this sentiment stated in one of the final scenes of the movie where Kyle's friend, Cartman, tries to bond with Kyle as they face potential death:

> CARTMAN [hunkering down, with Kyle, in the trench]: Kyle? All those times I said you were a big, dumb Jew? I didn't mean it. You're not a Jew.
> KYLE: Yes, I am. I am a Jew, Cartman!
> CARTMAN: No, no, Kyle. Don't be so hard on yourself.

The anxiety that causes this joke to be funny is the idea that being Jewish is inherently bad and a constant source of self-hatred; consequently, Cartman can only bond with his friend by trying to tell him to not take his Jewish self-hatred so seriously. What may tie this question of Jewish self-hatred to the Holocaust is the idea that Jews hate themselves for being victims, and their victim status comes from the Holocaust and popular culture.

This topic of ethnic self-hatred is one of the most difficult issues for students to consider. Since so much of popular culture seems to be about self-love, it does not make sense to many students to think about the roles played by self-negation in our society. Yet, I would posit that this type of self-conflict is exactly what needs to be explored in a class dedicated to positive social change because the inability of people to acknowledge the multiple parts of their own identities results in limiting self-knowledge and demonizing others who are connected with the rejected parts of the self.[15]

On Jewish Self-Hatred

In order to further explore this question of ethnic self-hatred, I often refer my students to Sandor Gilman's work on anti-Semitism. In his book *Jewish Self-Hatred*, Gilman argues that a central driving force behind internalized anti-Semitism is the Jewish desire to assimilate into the dominant culture. He posits that in order for Jewish people to fit into the society in which they live, they must accept not only the values and mores of the dominant reference group but also that group's fantasies about Jews (2). Yet, the dominant group sends a double message to Jews: you should be like us, and thus lose all your ethnic differences, and you will never be like us because of the differences we have attached to you (2). Assimilation therefore offers a double bind for the Jewish person and all other minority groups, and Gilman posits that this double bind is then internalized (2–3). In this sense, self-hatred is based on the minority group's identification with the dominant group's hatred of any type of ethnic difference.

Gilman adds to this argument the idea that the assimilated person always knows that he or she has not completely assimilated, and thus there is always a lingering sense of failure and rejection that is either internalized or projected onto other members of one's own group. For instance, Gilman argues that most of the Jewish jokes Freud discusses in his theory of humor are based on demonizing Eastern European Jews (264). Gilman also posits that Freud makes fun of these other Jews so that he can split off the good Jew from the bad Jew and differentiate between his own assimilation as a good Austrian scientist and the failure of other (Eastern European) Jews to assimilate. Moreover, Gilman thinks that one reason why Freud wanted to do a scientific study of Jewish jokes was that Freud believed science offered a universal non-Jewish language (268). Freud's turn to "objective science" is therefore itself an attempt at assimilating into a universalizing discourse; however, this act of assimilation constantly fails, and so he must perpetually return to the question of Jewish identity.

The irony of Freud's attempt to create a non-Jewish universal science of the mind was made apparent when the Nazis and other people labeled psychoanalysis the "Jewish Science." From one perspective, this term is accurate because the question of Jewish identity has often haunted the ability of psychoanalysis to be considered a modern universal science of the mind. For, Judaism has always represented a challenge to universalizing discourses, whether they are the universalizing discourse of Christianity, American assimilation, modern science, or Fascism. Not only do the Jewish people often consider themselves to be God's chosen people, they have also been labeled as different from the dominant cultures in which they live. Jewish particularity has therefore often come into conflict with the universalizing tendencies of Western culture. Yet, in this age of identity politics, one would think that this particularizing ethnic identity formation would be a source of power and self-affirmation. However, as my analysis of *South Park* shows, ethnic identity has become a major form of debasement and entertainment for the general public.

In order to understand why Jewishness and other forms of ethnic identity represent sources of humiliation and self-hatred in culture today, students can take into account the link between the universalizing global economy and the popular culture of assimilation. Since capitalism, like modern science, is ideally a universal discourse that treats every participant in the same way, it tends to take on an air of being tolerant of everyone and everything. As Jean-Paul Sartre argues, a price theoretically does not change when a Jew or a Muslim approaches the price tag, just like a scientific experiment does not change if a Jew or a non-Jew does it.[16] Jews and other minority groups thus turn to science and capitalism in order to escape their own particular identity and to enter into a discourse where they can be treated as equals. This equality is the ideal of modern universal tolerance: everyone should be treated equally, regardless of race, creed, or gender. However, postmodern culture turns this modern ideology on its head, and instead of preaching universal tolerance, it often preaches universal intolerance. Thus, what becomes circulated and distributed is not some purely abstract science experiment or price tag but a system of prejudices and generalizations in the form of the mass media.

South Park AND THE UNIVERSAL WORLD WIDE WEB

I have found that one of the best ways to help students come to terms with the global universalizing tendencies of cultural assimilation is to have them research on the Web the movies we watch in class. We should not dismiss the opinions of people on the Internet as coming from extremists or crazy loners: in many instances, Web-based discussions are very close to how students and other popular audiences receive media. For example, in the epinions.com discussion of the *South Park* movie, my students found many messages proclaiming the importance of free speech and meaningless entertainment and the right to spout prejudices. In the following example, one can see how this movie is passionately defended:

> Let's face it, humor is all about being surprised. And what is more surprising than watching a sweet innocent little child vomit out filthy language that would make a two-dollar hooker blush? The plot itself is irrelevant and the clever songs are unimportant. If you want to be a little embarrassed because you think people making music with their explosive intestinal gases and then lighting them is funny, then indulge. So rent South Park, and for the love of all that's holy, don't try it at home. (Note: the use of a French phrase in the text of this review in no way implies anything about the sexuality of the reviewer nor will it ever ever happen again so long as he lives.)

While this Web reviewer claims that the movie's plot is irrelevant and the songs are unimportant, at the same time he asserts that what makes this film worthwhile is its shocking quality. But, we must ask, what is so shocking about bathroom humor and racism? Moreover, we must question why this writer has to defend his own sexuality. Did the homophobia in the film touch on his own homophobic feelings? Furthermore, doesn't the phrase "two-dollar hooker" illustrate the way that a movie like this fuels people's prior prejudices?

Against this background of recirculated prejudices, many other responses center on the question of free speech:

> Putting aside the basic message of the film—that freedom of speech means EVERYBODY's speech, and not just that which you think is appropriate, and its urgent plea that parents pay attention to what their young ones are up to and talk to them regularly—not just going off half-cocked when they get in trouble, this film has enough in-jokes to keep the dedicated South Park viewer occupied for hours, and enough pop culture references to keep even someone who is utterly unfamiliar with Cartman and friends howling.

Here we find the defense of everyone's right to free expression contradicted by the desire to stop people from "going off half-cocked" and criticizing the media. As in many other responses, this writer's reaction implies that the circulation of hate speech is not the problem; rather, the issue is that some people are intolerant of other people's intolerance.

One way that free expression is defended by people on another Web discussion list is by attacking the film rating board for being Nazis. They write, "Be glad that the filmmakers have made complete and utter fools out of the MPAA rating board and self-righteous hypocrites jumping on the 'blame violence on the media' bandwagon. And whatever you do, try your hardest to sneak into this movie. It's a movie that's a breath of fresh air, in a Hollywood that is having its balls handed to them by Washington Nazis. All that is needed is that you check that serious, PC side of you at the door, sit back and laugh like hell" (dream.magic.com). Here we see how a total lack of historical specificity allows one to equate a country bent on exterminating a group of people with an organization that is trying to regulate the amount of violence depicted in the media. Other Web discussants use the decontextualized term *Nazi* to attack anyone who preaches political correctness or who tries to defend the right of minorities not to be attacked in the media by viscous stereotypes. In these examples, the Holocaust is invoked in order to blame the victim and to remove National Socialism from its original historical context.

As many people on the Web proclaim, *South Park* is able to discuss important issues like anti-Semitism and the Holocaust in a forum where these issues are no longer taken seriously. The overall effect of this strategy is to undermine the possible lessons that people could have learned from a more thoughtful public discourse on these topics. This undermining of social discourse is evident in the following statement by a student: "It's making fun of issues that the country is dealing with today, racial prejudice, and stereotypes, religious differences and alien abductions. The show is making fun of itself. As college students we know not to take the show seriously."

Here we see how one of the most effective aspects of postmodern assimilation is the ability to get people to internalize the prejudices that affect them and others and then say that these prejudices do not matter. In the case of *South Park*, the stereotypical demonization of the Jewish mother and the feminized Jewish male is hidden beneath a cloak of universal intolerance and a strong claim that popular culture is really about nothing. However, we know from Freud's study of jokes that humor *should* be taken seriously and that entertainment *does* have an important meaning.

In fact, I would argue that one central aspect of rhetorical studies is a desire to examine all modes of public discourse and to challenge all universal claims. As Freud and Stanley Fish show, a history and an agenda lie behind every claim of universality. Like Freud, we must realize that the first step to the critical awareness of popular discourses is sensitivity to the ways that people resist analyzing these important aspects of culture. As my reading of *South Park* shows, the assimilation of symbolic references to the Holocaust and anti-Semitism often results in a decontextualized discourse in which idealized universal values of tolerance and free speech are proclaimed, while hate speech is circulated in a responsibility-free public forum.

Teaching the Rhetoric of Resistance

To help work against this circulation of hate speech and the demonization of tolerance, I often ask my students to apply the four modes of rhetorical resistance to their own experiences with assimilation. One effective writing assignment has them discuss the key moments in their life where they had to make an important choice about entering into a new identity group. I ask them to explain what was lost and what was gained by their assimilation into this subculture. They are asked to end their paper by exploring the emotional and intellectual difficulties they had writing their essay.

Part of the idea behind this assignment is to have students see that the knowledge from class discussions can be used to reflect on both affective and intellectual aspects of their development. Also, by concentrating on the rhetorical defenses, I prevent the paper from becoming a simple expressivist narrative. In fact, I have found that although many students do struggle writing this paper, they are able to appreciate the need to take course concepts seriously and use educational knowledge as a mode of critical self-knowledge.

In the next chapter, I switch my focus from having students learn about the Holocaust and the rhetoric of resistance through the analysis of popular media to a discussion of how the representations of the Holocaust and anti-Semitism are shaping college learning environments and our culture in general. Central to this work will be a discussion about how liberal and conservative writers and politicians are using the rhetoric of trauma to manipulate popular audiences and to undermine academic freedom.

Chapter 6

Teaching Against Binaries: Anti-Semitism and the Holocaust in the Culture of Rhetorical Reversals

As I highlighted in the last chapter, the type of critical teaching I am advocating has recently come under attack by what can be called a "rhetorical reversal." We now live in a cultural climate where teachers who examine intolerance are called intolerant, where victims of prejudice are often seen as victimizers, and where perpetrators of prejudice are represented as victims. These reversals have been fueled by a concerted political attempt to downsize the demonized welfare state and to cut taxes for the wealthiest people in the United States. Within this political context, in order for the most powerful people to claim that they are the ones who really need tax breaks and special subsidies, the most dominant groups in our country have had to claim their own victim status.[1] Furthermore, the main way that these tax cuts have been rationalized is by attacking the "welfare state" and the perceived special treatment of minorities in America. In this reversed world, minorities are represented rhetorically as victimizing the majority by making the wealthy pay for welfare programs. Here we see how one of the central problems of basing one's identity on a victim status is that it opens up the possibility for a rhetorical role reversal where the victimizer claims to be the victim, and the victims are seen as the victimizers.

Attached to this logic of victim reversal is the current strategy of different conservative groups to use the Web and other media to monitor and censor faculty under the banner of "academic freedom."[2] Not only have these cultural attacks functioned to silence many cultural critics and progressive teachers, they have also worked to polarize our society and educational institutions.[3] One of the guiding forces behind this rhetorical movement is the use of anti-Semitism and the Holocaust as a rhetorical method to control debate through fear and guilt. In this cultural climate, we find the very successful neoconservative effort to defend Israel and the United States' foreign policy by equating any criticism of Israel and America to be a sign of anti-Semitism and Holocaust denial. Furthermore, in order to ensure that teachers are towing this conservative line, many groups on the Right have used the Web and other media to demonize critical academics as being not only anti-Semitic and anti-Israel but also anti-American and pro-terrorist.[4] These conservative groups employ a universalizing and polarizing rhetoric, which claims that anyone who is not a total advocate of conservative political positions must be both a liberal and a Marxist, anti-American, pro-terrorist Communist dedicated to indoctrinating students with Leftist ideology.[5]

Pedagogical Strategies in the Age of Identity Politics

To help teach students how to escape from the polarizing rhetoric so evident in the recent culture wars, teachers need to employ several pedagogical strategies. One vital technique is for instructors to avoid mentioning the name of any political party or politician in order not to alienate students with divergent political identities. Although this process may be hard to do, it is necessary because students are quick to dismiss or accept any argument that can be attached to their own ideology or an ideology they reject. In fact, I have found that by attacking one party or another, teachers do not promote social change and critical thinking; rather, they only intensify self-justified idealized positions and the demonization of others. In order to break this vicious cycle of idealization and debasement, teachers can model for their students a nonbinary mode of analysis.[6]

Another way of trying to break out of the endless cycle of self-justification is to have students examine the political ideologies they reject. However, instead of asking students to simply support or deny the ideologies they analyze, students can be asked to see how these positions are shaped by rhetoric. In fact, one of the most successful writing projects I have had my students undertake is to listen to radio talk shows and to record and analyze the rhetoric of the opinions of the callers. This assignment helps students recognize the importance of popular views while motivating them to develop a more critical perspective on popular opinions and prejudices. Once again, the idea behind this pedagogical strategy is to use rhetoric as a way of moving students beyond the simple acceptance or denial of an ideology.

To keep students away from polarizing views, I ask them to avoid any universalized or idealized claims in their writing. By working against these rhetorical defense mechanisms, students are pushed to contextualize their own arguments and to think about the roles played by such mechanisms in common discourse. This move against universalizing and idealizing rhetoric is an important step toward a more nuanced way of perceiving the world. Nevertheless, it is difficult for students *and* teachers to give up these rhetorical mechanisms because defensive processes help people feel certain about their own positions. This is achieved by reinforcing one's identity through the self-righteous demonization of another's position(s).

Along with avoiding the binary logic that shapes so much of the rhetoric circulating in our culture, it is also essential to resist a universalized mode of tolerance and relativism that undermines the ability of people to take a position on any issue. Instead of going to the extreme of pure postmodern relativism, we thus need to affirm that although there is not always a single right answer to a social problem, there are many wrong answers. In other words, we must develop a pedagogy that affirms multiple causes of a social issue and is still able to reject wrong or misleading solutions. Ultimately, what we want our students to be able to do in order to be critical thinkers is to work through issues based on the individual merits and contexts of particular social and psychological conflicts.

Many educators have argued recently that students cannot perform this type of critical thinking because they do not have the historical awareness necessary to contextualize and historicize the rhetoric they are analyzing.[7] In response to that objection I propose that a historicized version of rhetorical analysis can offer teachers a strong method for motivating students to be critical thinkers invested in social change. In fact, one reason why I use the topic of the Holocaust in my classes is that this cultural trauma forces students to see social issues in a historical light. However, as Mark Bracher has warned us, this turn to history should not make us think we are escaping from the need to contextualize and critically examine how the past is used to serve particular rhetorical purposes.[8]

HISTORICIZING THE RHETORIC OF THE HOLOCAUST AND ANTI-SEMITISM IN AMERICAN POLITICS

One way I have tried to get my students to take a more historical and rhetorical view of the prejudice and cultural trauma in contemporary society is to have them read sections of Peter Novick's *The Holocaust in American Life*. Novick's central rhetorical argument is that since anti-Semitism no longer exists in America, Jewish people and Jewish organizations have to invent it.[9] In this rhetorical reversal, students see how language is used to blame the victims of prejudice for the problems of prejudice. The first step in this rhetorical movement, which I believe we need to reveal to our students, is the idealized claim of universal tolerance and innocence so common in American culture and politics. In fact, without this rhetorical move to deny the presence of discrimination in our culture, it becomes much more difficult for people to claim that the victims of prejudice are the ones who really invent prejudice.

According to Novick, the major way that powerful Jewish groups create anti-Semitism is by reminding people of the horrors that Jewish people suffered during the Holocaust. Moreover, the main reason why Jewish organizations create this myth of anti-Semitism is to help raise funds and political support for Israel. One thing I want my students to understand in relation to this type of identity-based political rhetoric is that it is often centered on a denial of the multiple possible avenues for personal and ethnic identity.[10] Thus, instead of seeing identity as a result of competing and conflicting cultural, personal, historical, traditional, and social influences, the political manipulators of identity tend to trace all aspects of identification to a single cause of traumatic victimization. For example, in order for Novick to make his argument, he first starts with the questionable claim that the *only* current source of identity for assimilated Jews in America is the Holocaust and anti-Semitism (7). This universalized, identity-based theory is then followed by two central ideas: one, anti-Semitism no longer exists in the United States, and two, American Jews are one of the most privileged groups in the country (9).[11] Therefore, according to Novick's logic, since Jews can base their identity solely on anti-Semitism, and anti-Semitism no longer exists, Jewish people must turn to history to make their claim for victim status. The rhetorical trick that students should be able to detect here is the placing of ethnic identity in an impossible position: On the one hand, ethnic identity is seen as the only possible source for Jewish identification; on the other hand, its existence is denied.

One short writing assignment teachers can use to expand this type of analysis is to ask students to listen to poplar radio shows or examine political Web sites to see whether other social and ethnic groups are represented through this type of self-consuming rhetoric.[12] This kind of research is beneficial because it shows students that Jewish people are not the only group facing this new form of prejudice. Teachers must reveal how structures of prejudice affecting diverse social groups are both the same and different.[13] For example, students can be asked to compare the way homophobia is represented in the popular media vs. how anti-Semitism is circulated. By comparing and contrasting diverse modes of prejudice, students move away from the destructive notion that Jewish people are always victims or that Jewish identity can come only from the Jews' relation to the Holocaust and anti-Semitism.

This need to compare and contrast the different prejudices in our society is necessary in an academic setting because so many academic writers reduce ethnic and personal identity to a question of victim status and victim competition. For instance, the notion of victim status plays a large role in Novick's own assimilated anti-Semitism.[14] According to his reductive and universalizing theory of victim competition, it became popular after the Civil Rights movement for each ethnic and sexual group to base its identity and difference on a founding historical trauma. Thus, the African-Americans have slavery, the Native Americans have the American genocide, and the Jews have the Holocaust. Novick sees this process as reflective of a cultural shift from the celebration of heroes to the celebration of anti-heroes (8).

It is important to discuss with students this type of universalizing rhetoric of identification, which posits that the best way to get attention and power in our current culture is to stress the way your particular group has been victimized. On one level, students need to see how this popular notion of victimization can function to deny the real issues and problems facing a particular group, and on another level, the use of the victim label often results in a quick reversal where the victim is attacked for being a victimizer. One way of avoiding this type of labeling and reversal is to stop using the victim vs. perpetrator paradigm and to stress specific complex social dynamics. However, I am not arguing that we should simply equate victims and perpetrators; rather, I believe it is necessary to see how other roles complicate these situations.[15]

For example, Novick posits that since Jewish people in America are represented as being very powerful and wealthy, it is now hard to stress their victim status. Instead of simply dismissing his argument, however, it may be important to first acknowledge the difficult situation all "minorities" are placed in within American culture. In other terms, a more nuanced and accurate analysis would look at the multiple and conflicting messages that certain "successful" minority people have to encounter. In fact, many minority students now in college have often not only witnessed respect from their own group and the dominant majority but also experienced rejection from their own group for either idealizing the majority identity or trying to give up their own identity. On a psychological and rhetorical level, identity formation is therefore never a simple matter of placing oneself on one side of a binary opposition; instead, we need to see how all identity positions are conflicting and multiple. Yet, academics like Novick resist this type of a more nuanced mode of identity analysis. They seem bent on attacking their own culture while simultaneously claiming that their people are privileged and therefore never under attack.

Novick's reductive and universalizing internalized anti-Semitism comes to the forefront when he posits that Jews can gain "moral capital" only by turning to the Holocaust (9). I find his use of the term *capital* in this context to be highly questionable, and his rhetoric indicates that he will go to any lengths to deny both the importance of learning about the Holocaust and the presence of anti-Semitism in our present world. When examining this type of argument, teachers must ask students why someone would seek to ridicule and mock members of his or her own group to such an extent. One way of responding to this question is to point out how each group is structured by multiple identity conflicts that are often hidden by a false sense of group unity. In turn, individual members of a social group internalize the basic identity conflicts on an unconscious level, then these individuals identify with one part of the conflict and attack others for identifying with the other part. For example, at the center of Novick's whole attack on Jewish identity is his criticism of Jewish people who try to convince the American government and other Jews to support the state of Israel (149–69). Instead of acknowledging that many individual Jews in America are divided on this issue, Novick falls into the trap of detecting a vast Jewish conspiracy, and he projects his aggression onto the Jewish organizations he sees as leading this conspiracy.

One interesting historical aspect to point out to students about Novick's argument here is that the dominant cultural position has been reversed. That is, we are now seeing conservative groups endorsing Israel and labeling any criticism of Israel as anti-Semitic. Moreover, the conservative attack on victim politics has also been inverted so that the powerful conservative majority now claims to be the true victims of American society. Yet, using the term *conservative* in class will only alienate students and allow them to simply stay comfortable in their preestablished political affiliations. In order to avoid this potential conflict, a teacher can simply replace the word *conservative* with backlash or rhetorical reversal. This substitution of terms can also help move the analysis away from political identities toward more social and cultural mechanisms of identification.

A Universalizing Take on the Holocaust

One way I avoid the trap of political identification is by showing students how writers like Novick rely on the manipulation of rhetoric to make their misleading arguments. For instance, one of Novick's central claims is that Jews must cite the Holocaust as unique in order to secure their own special place as victims and chosen people. He also posits, however, that every historical event is unique, and thus no event is unique (9). What I try to reveal to my students here is how his rhetorical strategy is to take an argument based on difference and universalize it to the point that difference becomes impossible. Therefore, like the universal discourse of tolerance, this mode of argument can function to efface the possibility of ethnic and cultural differences.[16] Once again, I emphasize in class that the analysis of rhetorical defense mechanisms allows us to move away from a polarized political debate by concentrating on the methods used to construct false arguments. Moreover, this type of rhetorical analysis combines an attention to historical specificity with an awareness of how individuals internalize these larger social mechanisms. Ultimately, we can prove that Novick's take on the Holocaust and anti-Semitism is self-contradictory and that these contradictions arise because he is not so interested in analyzing a problem in a rational way. Rather, his goal is to utilize preexisting defense mechanisms in order to undercut the possibility of making any positive pedagogical or social contribution.

A common rhetorical move that Novick and others employ to undermine pedagogical efforts to use the Holocaust as a learning experience is to argue that every and any lesson has been drawn from this historical trauma, and therefore no lessons are appropriate (12). To makes this universalizing, antieducational point, he posits that the Holocaust has been employed for anti-American purposes by exposing the failure of the Americans to save Jewish people (12). Yet, he also indicates that the Holocaust has been invoked to glorify the Americans as the great liberators (13). Instead of examining the possibility that both of these historical claims may be true, his polarizing rhetoric forces him to posit that *every* major cause and victim group has used the example of the Holocaust to point to the importance of their particular political program. From this universalizing perspective, it becomes easy for him to claim that the Holocaust is too extreme to provide any type of example or lesson (13). What I want my students to learn from this type of self-consuming rhetoric is the way that Novick first denies the uniqueness of this event but then claims it is so unique and extreme that nothing can be learned from it. A careful analysis of this rhetoric shows students the need to detect contradictions in arguments and to look behind rhetorical ploys for the underlying positions.

Thus, even though Novick claims that we can learn nothing from the Holocaust, he still writes a lengthy book on the subject. Why does he do this, and why would he pursue such a subtle form of Holocaust denial? One of his central goals appears to be to undercut the United States' policy of supporting Israel. While I do not think that criticism of Israel should be equated with anti-Semitism, I do think it is important to see how prejudice and trauma are used in the rhetorical construction of political positions.[17] For instance, one of

Novick's central claims is that the support for Israel must be tied to the Holocaust in order for Jews to hold a position of moral righteousness and to avoid any level of criticism (10). According to this rhetoric of empathic identification, by forcing people to remember the Holocaust, American Jews place themselves in an idealized position where every one of their actions and beliefs becomes justified (10). Moreover, by equating the victims of the Holocaust with the citizens of Israel, Novick posits that American Jews are able to make a strong moral claim for support (10).[18]

I do think it is important to question aspects of Israel's politics and the role of the United States in supporting Israel. Novick, however, takes these important concerns and wraps them in a rhetoric that both attacks Jewish people and undermines the important lessons of the Holocaust. For example, many of his claims center on the stereotypical idea that American Jews are highly defensive people who cannot tolerate any sense of criticism and who need to be seen as victims.[19] He also returns to the prejudicial idea that Jewish Americans are self-righteous people who are blind to the suffering of others. Furthermore, he states that Jews are inherently suspicious of Gentiles (180). It appears that Novick is intent on blaming the victim of anti-Semitism while he claims that the victim does not exist in the first place. Reversing Jean-Paul Sartre's dictum, Novick posits that it is the Jew who invents the anti-Semite.

One of the ways that this Jewish invention of anti-Semitism is supposed to occur is through the Jewish control of the media (12). In this updated version of the classic anti-Semitic myth, the powerful Jews are able to manipulate other people by getting them to feel guilty through the constant media representation of the Holocaust. This argument not only rests on the false universalizing claim that the Jews control the media but also repeats the current idea that the Holocaust has been overrepresented. By pointing to a few movies and one television series, Novick posits that everyone knows about this event, and they are tired of hearing about it. Furthermore, he believes that these popular representations teach us nothing and are ineffective as learning devises (212).

In opposition to Novick's desire to refute the importance of popular culture and the circulation of prejudices in our society, I have argued that we must take popular media seriously and look at the different defense mechanisms through which stereotypes and intolerance are recirculated. Thus, instead of dismissing the presence of anti-Semitism in everyday life, we need to expose and critique it in an open and thoughtful way. However, even though we may stress the idea that anti-Semitism still exists in the world, we still have to dissociate this prejudice from the foundations of Jewish identity.

Certain modes of cultural and ethnic studies, however, confuse the goal of teaching historical traumas with the need to support students' quests for a solidified identity. Thus, Novick is partially right in his criticism of identity politics and the reductive, universalized idea that one should base one's ethnic identity on a founding trauma. Yet, in his dismissal of this type of pedagogy he goes too far, and he attacks the whole idea of learning about historical traumas.

Moreover, like so many other contemporary academics and students, Novick's universalizing rhetoric pushes him to fall into the postmodern trap of claiming that since history is often perceived through fictional reconstructions, history must always be a fictional representation. In this vein, Novick posits that the Holocaust has now become based on a "retrospective construction" that turns history into mythology (20).[20] According to Novick's own account, one of the reasons for this transformation of the history of the Holocaust into the mythology of the Holocaust is the Jewish desire to fight off the effects of American assimilation by highlighting the mythic aspects of Jewish suffering. Moreover, part of this argument is based on the fact that due to the high rate of intermarriage and the low rate of religious participation, the Jewish population in America is quickly declining. In order to counter this trend, Novick claims that Jewish organizations employ the myth of the Holocaust to help Jewish people gain a sense of ethnic identity and stop the destructive tide of assimilation (171). For Novick, this strategy represents a shift from the Jewish desire just to fit into American culture to a more inward turn toward Jewish ethnic identity (171). Moreover, Novick falsely equates ethnic difference with victimhood status, and this false identification allows him to describe a vast Jewish conspiracy of getting Americans to feel guilty about the poor plight of successful American Jews.[21] According to Novick's logic, the main tool that these Jewish organizations use to manipulate the American public is the popular representation of the Holocaust (171).

Novick is therefore forced to contradict himself: namely, he asserts that the popular culture depiction of the Holocaust has no effect on people *and* that the Jewish use of popular culture to represent the Holocaust has a great effect on the general populace. His argument not only rests on the repetition of the anti-Semitic notion of a Jewish conspiracy but also plays off of the idea that the Jews are both too powerful and not powerful at all. This contradictory notion is reflected in his references to popular culture where he constantly mocks and trivializes the circulation of anti-Semitism. For instance, he labels the anti-Semitic remarks of militant blacks as "laughably trivial" (172). Yet, he then goes on to state that Jewish organizations want to make fellow Jews feel anxious by constantly invoking reference to anti-Semitism in popular culture (176). According to Novick, these organizations create fear in their Jewish constituents by proclaiming that there is a natural and inevitable movement from anti-Semitic jokes to concentration camps (178). What he leaves out of this exaggerated argument is the space between anti-Semitism and the Holocaust. Yet, he cannot point to this space because he would have to face the fact that anti-Semitism is still alive and well in American popular culture and life.

One of the most extreme examples of Novick's own internalization of anti-Semitic rhetoric is his claim that the only reason why Holocaust museums exist in America is because Jewish people exploit their wealth and political power (195). Once again, this argument relies on the anti-Semitic theory that Jews secretly control the world through their manipulation of politics and money. We know that the Nazis employed this same idea in order to convince the German people and themselves that the small Jewish minority (1 percent of the German population at the start of World War II) was actually responsible for the failures of the German economy and the threat of Communism. In this rhetorical structure, the Jews were positioned to be both hyper-capitalists and hyper-socialists. Moreover, they were seen at the same time to be a weak victim group and a powerful political organization. Novick employs this same faulty logic by insisting that Jews are winning the victim competition through their powerful control of politics and the media.

As teachers, we must help students see that Novick's use of this notion of victim competition appears to be derived from a political attempt to reject the legitimate claims that various minority groups have made in the American political arena. In this rhetorical reversal, he mocks minority groups by arguing that the stress in the media on inner-city poverty, racism, spousal and child abuse, and homelessness has shifted the focus of American culture from winners to losers (190). Through this rhetoric, he implies that Holocaust victims and survivors are simply losers whose only victory comes from their power to control the media and make the world feel guilty.[22] One reason why Novick can make this equation between victims and losers is that he refuses to examine the multiple identities and histories of these minority groups, including American Jews. In fact, Novick mocks the "third-generation renaissance" of ethnic consciousness as a purely "symbolic" and "undemanding" return to Jewish rituals (186).

Novick's internalization of anti-Semitism is coupled with this extended effort to rid every minority group of any type of complicated identity. He also appears to be bent on denying the significance of the Holocaust and the presence of anti-Semitism in American culture. One would think that all of these negative arguments would be linked to a conclusion stating some type of positive program or claim regarding the Holocaust. At the end of the book, however, the reader is left only with the reiterated argument that everyone now knows about the Holocaust, and there is really nothing to learn from it. We are also told that the Jewish claim of anti-Semitism in our current culture is "nonsense" and that the Holocaust represents a "black hole" for theology (269). However, doesn't Novick's own book represent a theoretical black hole that collapses every effort of enlightenment into a condensed space of nothingness? This effort to reach a universal state of non-meaning represents one of the strongest aspects of American assimilation. For, what often hides the internalization and circulation of racist and anti-Semitic cultural stereotype is the universal claim that culture and history have no real meaning.[23] Yet this universal nothingness is most often employed to hide a particular political and/or personal agenda.

I think it is essential to show to students how Novick's rhetorical agenda is an attempt to undermine the claims made by all minority groups for some type of recognition in the general American public. In the case of Jewish Americans, Novick wants to deny them the status of a minority group in order to blame them for their own suffering. Is this not a classic case of internalized anti-Semitism hidden by a universal rhetoric of non-meaning?

Students Analyzing the Rhetoric of the Holocaust and Anti-Semitism after 9/11

While it may be easy to dismiss Novick's arguments as harmless academic posturing, many of his central claims have been reiterated in the popular media and in academic research. In fact, I argue that a careful analysis of such arguments circulating on the Web can provide teachers with an important method of applying the critical analysis of rhetorical defenses to an understanding of the continued role that the Holocaust and anti-Semitism play in shaping not only American politics and culture but the global political world as well. To help my students analyze this political rhetoric, I have them research Web sites and radio talk shows that present a strong political view.[24] The first part of this research asks students to examine the different rhetorical devices that are used to construct political arguments. I ask students, though, to avoid deciding whether the content of the claims is right or wrong. My goal is not to simply push for a value-neutral model of tolerance; instead, I want students to move out of the polarized world of extremist political ideology by concentrating on how rhetorical figures make these political positions possible. For example, in a recent class one student analyzed an online letter written by David Horowitz, which touches on many of the issues I have been discussing throughout this chapter. To help clarify what is at stake in this type of critical research, I discuss several claims that Horowitz makes and how my student analyzed these claims in terms of political rhetoric. I also indicate how I discussed these potentially polarizing topics in my class.

One of the central rhetorical figures students frequently detect in strong political views is the use of universalization. The student analyzed the following quotation from Horowitz's online article "Help Stop the Anti-Israeli Divestment Campaign": "The anti-American left, which has formed a fifth column in this country to betray us from within, is also targeting Israel as America's ally in the Middle East. Israel is a tiny country and vulnerable. The left is testing its strength on Israel, but its ultimate target is the United States." This student pointed out that the letter starts by universalizing the Left and identifying Israel as the victim of a concerted attack. The student then went on to show how the attacks on Israel are equated with a generalized attack on the United States. While this student did concentrate on the rhetorical construction of this political argument, his analysis still introduced potentially divisive material. In fact, in one of my classes that had several strong supporters of Israel, it was hard to have an open and rational discussion of these issues. The technique I used in this course was to examine the arguments on a rhetorical level; if people tried to pull others into a political debate, I asked students to stop the discussion and start to free write about how they were feeling at that moment.

Many times these writing experiments allowed me to access conflicting ideas and emotions that students were having about the class content and the general subject matter. Sometimes I would just note these concerns to myself, and other times I would discuss with the class general trends in students' reactions. One of the benefits of this type of pedagogical technique is that it motivates students to confront their own conflicting desires and fears in a safe environment. For instance, students have written that they are afraid of voicing their political opinions to the class because they don't want to stand out. Other students have written that they do not like to discuss politics with their friends because it makes people uncomfortable. Still others have written that people see you as a "nerd" if you are too passionate about a political position.

It is important to analyze in class these types of subjective feelings and ideas because they are often excluded from the presentation of political ideology. The following claim by Horowitz that my student examined, for example, reveals that many extreme viewpoints rely on the subtle and not-so-subtle rhetoric alike of "guilt by association":

> What happens to Israel will eventually happen to America itself. Yasser Arafat built the first terrorist training camps. Arafat was Saddam's staunchest ally in the first Gulf War. Palestinian terrorists were involved in the first bombing of the World Trade Center in February 1993 and the destruction of the Khobar Towers barracks where 19 U.S. servicemen died in 1998. Suicide bombing began as a tactic of leftwing terrorists against Israel. It reached its culmination on 9/11.

One of the things my student pointed out about these statements is that they use real facts to convince the reader to believe in the argument, but these real facts are combined in such a way as to present false information. In fact, the central rhetorical movement that is repeated in so much extremist rhetoric on the Web and radio talk shows is to universalize both the victims and the perpetrators of political violence. Here, America and Israel are identified as the victims, while the Palestinians are equated with Saddam Hussein and the terrorist bombers of the World Trade Center.

While I believe that teachers should correct blatant misinformation in the content students analyze, I also believe that any direct attempt to simply clarify a position will often end up polarizing students' views. A more effective pedagogical strategy would be to respond to the previous statements by first examining the rhetorical movements that allow for the equating of U.S. and Israeli interests. An even better method is to ask students to trace the rhetorical movements that help produce arguments. Another important technique is to ask students to locate the point where the displacement of political terms occurs. For instance, in the passage quoted, Horowitz's term "leftwing terrorists" provides the key to his rhetorical strategy. In response to this type of slippage, I would simply ask students what the term means and why the writer employs it.

Teachers shy away from directly addressing these types of "extremist" arguments at some risk because these ideas are increasingly entering into our classrooms and our students' subjectivities. In fact, one of Horowitz's main missions is to intimidate teachers so that they do not critique America or Israel. Thus, in the following passage, we see how these political issues have been brought to our colleges and universities: "These America- and Israel-haters continually refer to Israel's "apartheid" policy, in order to evoke the same passionate hatred for Israel that college students and faculty used to hold towards the white government of South Africa." My student's response to this statement was twofold: he did not know what Horowitz was talking about, and although he had heard of the idea of divestment, he did not know what this issue had to do with South Africa or teachers. This student's response points to the fact that our students often have a limited sense of current affairs and recent political movements. Thus, when they read these strong political arguments, they are often left confused and alienated.

While most teachers would react to students who don't know the recent history behind a political argument by filling in the facts and details, I have found that this technique does not always work. This is because the problem has more to do with the subjective reaction to politics than the actual cognitive knowledge that students do or do not have. Unfortunately, we are trained as teachers to believe that knowledge is the solution to most problems, and therefore we tend to neglect the more emotional and rhetorical foundations of learning.

Thus, instead of simply telling our students what the correct information is, it is often more effective to have them research the missing historical contexts themselves because students may simply memorize or reject the information that comes from teachers, especially when that information deals with contentious political issues.

One way I have responded to statements like the previous one is to ask students to reflect on the moments when they felt a teacher was trying to shove a political ideology down their throats. After all, Horowitz's central complaint is that tenured radicals are indoctrinating our students to hate America, Israel, and Western civilization. It is therefore important for us to determine how students are feeling about their teachers' promotion of politics in the classroom.

A method I have used to study this question of political indoctrination is to have students distribute and analyze surveys they have their friends and classmates fill out. One of the questions on this survey asks students to discuss any instance when they felt a teacher was forcing a political belief on them. The most common response to this question is that although students may feel that their teacher has a strong political bias, that does not affect how the student will see the world. In fact, students report that since they are often graded on what the teacher thinks, they learn quickly how to give the teachers what they want without having to understand or believe the content of the information. I take these responses to indicate that even when teachers try to present their political ideas to students, this effort often backfires because students see this ideology as just another piece of information to memorize and then later forget. Teachers who therefore try to teach students certain political beliefs neglect the possibility that the subjective aspects of learning undermine the ability of teachers to simply shape their students' ideologies.

Not only do teachers often fail to see the difficulty of influencing their students' beliefs but also rhetorical manipulators like Horowitz often fail to understand how students tend to process information they learn in their classes. For instance, Horowitz makes the following claim: "It's at colleges and universities, above all, where the lies about Israel are finding receptive audiences. Many faculty, instead of seeking objective truth as their profession demands, have signed on to the radical left's agenda. They influence impressionable students with their hatred of the United States, Israel, and Western civilization." My student reacted to this statement saying that teachers do not influence students so easily and that this "Horowitz guy" treats students as if they were just one massive sponge. In addition, many other students have told me that they have never encountered any anti-American sentiment on campus. In fact, they wonder why these critics are attacking education when it is the politicians who are fighting all the wars.

Many students and teachers tend to see this type of political extremism as just the ranting of a few crazies on the Web, and they do not understand why we should take the time to study them in a serious way. My response is that these views are more common than we may think, and they do help to shape political legislation that may, in turn, shape the way teachers teach and students learn in the future. However, even this type of corrective information coming from a teacher does not always cause students to think critically about the topic under discussion. Consequently, I still find it necessary to provide ways for students to come up with their own research and arguments to analyze these sensitive political issues in an effective manner.

Social Issues Writing Assignments

One writing assignment that can be useful in motivating students to apply their knowledge of rhetoric to a social issue is to have them write research papers and then editorials on a social issue of their choice. I have found that by telling students they will have to write an editorial for the school paper, they take the research aspect more seriously because they don't want to look uninformed in front of their fellow students. Also, students often are more rhetorically aware of their audience and their own writing voice when they have to write for a real audience. In fact, I ask students in this assignment to anticipate in their editorials the defensive reactions of other students.

I require students to come up with a clear position on their topic, but they must resist using universalizing and polarizing rhetoric. One of the ideas behind this type of writing is that students need to constantly think about the role of rhetoric and various defense mechanisms in the construction of political arguments. In addition, although students are not required to actually publish their editorials, the mere possibility of having other students take their work seriously can push them to make more nuanced and careful arguments.

I also have students write a follow-up paper in which they reflect on the emotional and intellectual problems they encountered writing their research paper and editorial. One reason for asking them to discuss the emotional and intellectual aspects of writing—and the resistance to writing—is that I want to value the different aspects of their subjectivity equally. In fact, after I've read self-critical student essays I'm impressed with the way students bring up the social influences shaping their own ideas and writing processes. Thus, even though students are graded as individuals and usually work as individuals, their writing and thinking is often shaped by constant dialogues between what they want to say and what they think others want them to say. Although it is precisely this kind of inner conversation that is often repressed in most learning environments, it nonetheless still offers a key to the possibility of critical thinking and social change.

In the next and final chapter, I extend this analysis of the political and psychological context of teaching about positive social change in today's globalized world. One of my main concerns is the way that universalizing rhetoric has helped to create a "Global War on Terror" that has no apparent end or identified objective. Thus, this rhetoric has worked to construct a constant state of fear, which in turn is used to justify idealized acts of aggression and to undermine any attempts at critical analysis.

Conclusion

From the Holocaust to the Global War on Terror

Throughout this work I have concentrated on the interaction among displaced idealizations, traumatic victim identifications, the cultural assimilation of internalized prejudices, and the universal model of globalized indifference. To show how these rhetorical processes are relevant to contemporary society, I conclude this work by turning to a brief discussion of the roles of anti-Semitism and the Holocaust in the "Global War on Terror." As I discussed in the last chapter, the invocations of the Holocaust and anti-Semitism have been used to justify and idealize the United States' foreign policy and America's support for Israel. Thus, from this idealization perspective, we must support Israel because failing to do so will cause an increase in global anti-Semitism and could lead to another Holocaust. Note, in addition, that Islamic fundamentalists are using anti-Semitism to unify their people and to claim their own victim identification status. Moreover, these mutually reinforcing politics of victimization have become more vexed due to 9/11 and the ensuing Global War on Terror. For, after 9/11, the strongest nation in the world began to claim its own trauma and victim status, and this identification was soon employed to justify a series of idealized national aggressions.

The movement from victim identity to idealization is therefore one of the major sources of destructiveness in the world, and making matters worse is the global media system that uses popular culture to turn every real event into a fictional construction. In many ways the popular reception of the Holocaust I have traced throughout this book has provided the foundation for this new global order. I have shown that what often fuels the Global War on Terror is the use of universalized media to circulate trauma and prejudice. Simultaneously, the universalized media and political order claim to be neutral and tolerant. In fact, by showing that the traumatic nature of the Holocaust could be turned into an object of popular culture, the American system of globalized entertainment and assimilation was able to provide the rhetorical foundations for a disconnection of historical trauma from critical analysis. In turn, once this fundamental trauma was removed from serious thought, it then became an easy object of political and psychological manipulation.

Linked to this popularization of the Holocaust, we have found the emergence of a postmodern form of anti-Semitism and prejudice, which I related to assimilation and internalized racism. The key factor to this system is that people begin to repeat destructive prejudices about their own identity groups in order to become accepted by the dominant culture. Furthermore, this symbolic system of internalized prejudice is able to circulate hate while people claim a universal ideology of tolerance. Making matters worse, dominant groups are denying the victim status of minority groups at the same time the people in power claim that the dominant majority are the true victims. Thus, according to this backlash rhetoric, we should not only get rid of affirmative action and the welfare state but also give tax cuts and tax breaks to the wealthy.

The conservative claim for victim status has also entered into the realm of higher education and the Global War on Terror. Led by David Horowitz, Daniel Pipes, and a large number of conservative think tanks, there is a growing movement to target liberal professors for being anti-American and anti-Semitic. For example, in his book *The Professors: The 101 Most Dangerous Academics in America*, Horowitz criticizes faculty members who do not support the policies of Israel and the United States. According to Horowitz's universalizing rhetoric, the tenured radicals who control American higher education are bent on indoctrinating their students into a Communistic, anti-American belief system. Moreover, he labels any criticism of Israel or any support for Palestinians "anti-Semitic." Likewise, Horowitz condemns any academic who does not support all of the policies of George W. Bush and labels these dissenting academics "Marxist anti-Americans." Horowitz's central activity is to campaign for what he calls an "Academic Bill of Rights," which claims to be a political movement in support of the free speech of all students. Playing on the universalizing doctrine of academic freedom, Horowitz posits that leftwing faculty members who dominate higher education are oppressing conservative students. In order to have a more "balanced" approach to instruction, he calls for the hiring of conservative faculty and the policing of Leftist professors who stray from their subject matter and attempt to present their political opinions in class. While he has successfully gotten his bill to be considered in many state legislatures, it seems that his main objective is to censor faculty who are critical of Israel and America. Furthermore, a side effect of his attacks is to give ammunition to conservatives who want to defund public higher education.

As I have argued throughout this book, the key to this neoconservative rhetoric is the reversal of identity positions so that the real victims are seen as victimizers while the victimizers are represented as the idealized victims. Such neoconservative conversion narratives are becoming increasingly common in our world, and they point to the need for a psychoanalytic model of cultural criticism, which could effectively call into question all acts of political idealization and victim identification. In fact, by making our media more interactive and by giving up the easy notion of universal toleration, people may be able to come to terms with the conflicted nature of all identity systems. Part of this educational and cultural process will require a confrontation with the internalized anti-Semitism and racism circulating in popular culture. We therefore need to use a psychoanalytic model of rhetoric and pedagogy to challenge all political acts of idealization, identification, assimilation, and universalization. This is not to say that we can live without these important rhetorical mechanisms; rather, we should find ways to employ these tropes in a more complex and self-critical fashion.

What I have called for in this book is a psychoanalytical model of pedagogy that acknowledges the important roles played by rhetorical defense mechanisms in the construction and conception of popular texts. I have also posited that psychoanalysis shows us why we cannot simply attack these defense mechanisms head on; instead, we have to develop teaching strategies dedicated to providing safe spaces for self-critical analysis. A major aspect of this mode of instruction is an emphasis on the tight connection between student subjectivity and the possibility for teaching about positive social change. I have therefore argued that if teachers are really concerned about helping students become more active critical thinkers, they must find ways to recognize the diverse aspects of students' identities.

As I argued in the last chapter, an important step toward making teaching more socially responsible is to acknowledge that one cannot simply correct a student's lack of information by giving him or her new correct knowledge. In fact, throughout this work I have demonstrated that diverse educational structures undermine our ability to teach students what we want them to learn. For example, the stress on grades in higher education often blocks students from thinking critically about the information teachers are presenting. I have also depicted ways in which teachers fall into the trap of alienating students by presenting polarized and universalized views. Instead of simply trying to present students with our own values, we need to engage them in an interactive learning experience that avoids political labels and polarized debates.

While it may make teachers feel self-justified and confident to teach in a way that idealizes their own positions and demonizes the positions of others, I have found that this type of self-satisfied teaching rarely produces the intended results. It is therefore essential for us to develop pedagogical methods that avoid playing into the binary logic of idealization and demonization. One method I have stressed in this effort to change the way we teach is the use of nongraded, anonymous writing assignments and electronic discussions. Drawing from the psychoanalytic notion of free association, I have found that a careful use of nontraditional modes of communication can function to challenge the educational stalemates that block critical learning.

A central reason why I have emphasized the use of the Holocaust in the promotion of this critical mode of education is that this historical trauma represents for our culture the extreme results of what can happen if we do not teach against intolerance and social passivity. In other words, this book has been dedicated not only to keeping the memory of the Holocaust alive but also to presenting methods to help us learn from this tragic period of our cultural heritage. An aspect of this process has been to examine critically how popular culture has turned this historical trauma into a part of our entertainment industry. By asking what has a made the Holocaust popular, I hope I have started a much-needed educational dialogue.

NOTES

CHAPTER 1

1. For an insightful analysis of this backlash rhetoric against the teaching of tolerance, see John K. Wilson's *The Myth of Political Correctness*.
2. Dinesh D'Souza's *Illiberal Education* is a classic example of the rhetorical reversal, which details an attack on the teaching of tolerance and the promotion of intolerance.
3. A prime example of the conservative effort to demonize progressive teachers can be found in David Horowitz's *The Professors*. I discuss aspects of Horowitz's work in Chapter 6 of this book.
4. Some of the recent works that have examined the role of the Holocaust in contemporary American life and culture are Tim Cole's *Selling the Holocaust*, Norman Finkelstein's *The Holocaust Industry*, and Peter Novick's *The Holocaust in American Life*.
5. One of the rare exceptions to this failure of academic critics to take into account the diverse ways that audiences respond to the popular depiction of the Holocaust is Dominick LaCapra's *Representing the Holocaust* and *History and Memory After Auschwitz*. One thing I hope to add to LaCapra's important work is a more systematic analysis of the different defense mechanisms that shape the reception of historical trauma.
6. This notion of working through resistances extends LaCapra's use of transference in academic settings. See *Writing History, Writing Trauma* (106–8).
7. Aristotle's theory of rhetoric is one of the first efforts to combine a theory of individual psychology with an articulation of the role of language and discourse in public life.
8. In his *Risky Writing*, Berman advocates a mode of expressivist writing that is very effective at getting students to discuss difficult subject matter. I argue, however, that his nonjudgmental approach does not allow student defenses to be worked through.
9. While Bok is very good at exposing many of the problems of higher education, he—like many other educational critics—does not offer many possible solutions to the problems he elaborates.

10. Thus, in contrast to Berman's use of non-judged writing, I make sure that my assignments motivate the students to reflect on their own use of defense mechanisms.
11. Slavoj Zizek's repeated claim that the end of one's analysis involves an identification with one's symptom is a highly reductive and misleading reinterpretation of psychoanalysis.
12. See Nick Tingle's *Self-Development and College Writing* and Marshall Alcorn's *Changing the Subject in English Class* on this relationship between learning and mourning.
13. As LaCapra astutely points out, Zizek's use of the Lacanian Real and the psychoanalytic notion of trauma often repeat this empathic identification with the failures of representation (*History*, 45–48).
14. Freud argues that in the state of passionate love, it is easy for the lovers to regress to a state of criminality, and I would argue that we can use this same model to explain the regressive criminality of groups.
15. This stress on the inability of people to mourn loss in the structure of idealization is presented throughout Eric Santer's *Stranded Objects*.
16. See *Escape from Freedom*.
17. Throughout his work Lacan returns to this game to locate different possible interpretations. Some of his most important insights can be found in *The Four Fundamental Concepts of Psychoanalysis*.
18. See Descartes' *Discourse on Method* (41) for the development of scientific universality.

Chapter 2

1. One possible way of defining postmodernity is this new combination of education and entertainment. Thus, instead of these two aspects of modern culture being in conflict, postmodern society blends them together, and it is this combination that often undermines the academic desire for critical distance.
2. One of the few academic articles that takes into account how Holocaust museums actually try to educate people is Elizabeth Elsworth's, "The U.S. Holocaust Museum as a Scene of Pedagogical Address." Elsworth argues that the memorial is effective in challenging people's identities and notions of history. However, I believe this article relies more on academic wish fulfillment than on a critical analysis of how people actually react to the museum.
3. Some of the academic works I address throughout this work that tend to ignore the actual reactions of the public and of students are Tim Cole's *Selling the Holocaust*, Norman Finkelstein's *The Holocaust Industry*, and Peter Novick's *The Holocaust in American Life*.

4. For a helpful analysis of the limits of social science research concerning the study of prejudices, see Elisabeth Young-Bruehl's *The Anatomy of Prejudices*.
5. One can access Beckwith's text at: http://www.historyplace.com/pointsofview/beckwith.htm.
6. Gourevitch's article is located on the Web at: http://www.english.upenn.edu/~afilreis/Holocaust/gourevitch-museum.html.
7. On Piaget's notion of assimilation, see J. S. Atherton's *Learning and Teaching: Assimilation and Accommodation* (2005), available at: http://www.learningandteaching.info/learning/assimacc.htm.
8. Barbie Zelizer's *Remembering to Forget* offers an effective analysis of the depiction of the Jewish victims in photography.

Chapter 3

1. What is often missing in the cultural studies' interpretations of historical traumas is the role played by emotion and pedagogy in the experience and understanding of traumatic violence. In "Going Postal," Lynn Worsham defines emotion as "the tight braid of affect and judgment, socially and historically constructed and bodily lived, through which the symbolic takes hold of and binds the individual, in complex and contradictory ways, to the social order and its structure of meanings" (216). This theory of emotions helps us see why it is so difficult for us to engage our students in thinking about traumatic events, for the very structure of emotion entails a confrontation with the social construction of affect, and this encounter with social determinism is itself traumatic by nature. However, as Worsham rightly argues, we cannot ignore the important role that emotion plays in every pedagogical encounter (216).
2. Freud's neglected article, "Creative Writers and Day-Dreaming," outlines a properly psychoanalytic theory of cultural criticism.
3. For Spielberg's reflections on his religious experience during the making of *Schindler's List*, see Joseph McBride's *Steven Spielberg* and Michael Pascal's "Steven Spielberg."
4. In "Spielberg's Oscar," Omer Bartov discusses how Spielberg turns Schindler into an American hero (43) and a Christian Good Samaritan (48).
5. Yosefa Loshitzky analyzes Spielberg's depiction of Schindler as a Christ-figure (114) in his essay "Holocaust Others."
6. While many critics question Spielberg's commercialization of the Holocaust (Bartov, Gourevitch, Hartman), in "*Schindler's List* Is Not

Shoah," Miriam Bratu Hansen celebrates Spielberg's capitalistic aesthetics (97–98).
7. For Spielberg's own comments on the making of this film and its relation to his own Jewishness, see *Steven Spielberg*, edited by Lester D. Friedman and Brent Notbohm.
8. While it may seem that I am forcing students to accept my own interpretations, my goal is to offer a criticism of the defense mechanisms in the film so that students can later analyze their own defenses. One way that I try to prevent students from simply internalizing my interpretations is that I never grade them or test them on their ability to remember my analysis. I also want to stress here that while I do think it is important to allow students to express their own ideas, I still think that the teacher needs to model new ways of looking at cultural material.
9. What most critics do not understand about Lacan's notorious practice of ending his analytic sessions at variable times, not just at the classic fifty-minute mark, was that Lacan was trying to force his patients to talk so fast that they would not have the chance to censor themselves.
10. In "*Schindler's List* Is Not *Shoah*," Miriam Bratu Hansen discusses Spielberg's use of anti-Semitic stereotypes to depict the Jewish characters (83).
11. Whereas Schindler was given a golden ring, it was not done in the dramatic fashion outside the factory that the film depicts.
12. In "But Is It Good for the Jews?" Sarah Horowitz examines the representation of women in this film. She points out the way that the movie eroticizes the Jewish female victims of the Holocaust (127).
13. Some of the most troubling distinctions between the novel and the film surround the depictions of individual Jews and the efforts of organized resistance. In the novel, Leopold Pfefferberg represents a strong Jewish man who is constantly battling the system and trying to help his fellow Jews. The fact that he is virtually absent from Spielberg's version shows how this director needs to de-emphasize for dramatic effect the role of the Jews in their own liberation.
14. Cathy Caruth's important work *Unclaimed Experiences* posits that postmodern theories of subjectivity and representation, which problematize our notions of historical reference, may work to undermine our efforts to take an ethical stance in relation to traumatic experiences (10). However, Caruth affirms that the very nature of trauma replicates many aspects of postmodern culture and theory: "Trauma describes and overwhelming experience of sudden or catastrophic events in which the response to the event occurs in the often delayed, uncontrollable repetitive appearance of hallucinations and other intrusive phenomena" (11). In this structure, trauma pushes us to rethink our conceptions of history and reference so that we take into account the radical temporal distinctions between an event and its representation.

Moreover, this separation of the referent from the sign of the referent points to a postmodern and deconstructive notion of cultural representation.
15. In *The Vital Illusion*, Jean Baudrillard argues that contemporary culture uses the past in order to whitewash history (33–34).
16. In *The Plague of Fantasies*, Slavoj Zizek has articulated some of the ways that our cultural representations of trauma allow us the ability to watch other people suffer while we remain sitting in the comfort of our own homes. For example, in discussing the case of the traumatic war in Bosnia, Zizek argues that the West was able for a long time to stay out of the conflict by maintaining the presence of a "neutral gaze" (17). Furthermore, by not taking sides with any of the main parties, Zizek posits that the West tried to maintain "the impossible neutral gaze of someone who tries to exempt himself from his concrete historical existence" (18). I believe that this false neutrality represents one of the major forces defining the ways my students react to the Holocaust and other cultural traumas: in this structure, neutrality represents a repression of history and social determination.
Zizek not only sees this impossible neutral gaze as a mode of social and personal resistance to working through the cultural foundations of trauma but also argues in *The Metastases of Enjoyment* that trauma itself can be defined as the "gaze of the helpless other" (210). In other terms, what we find so threatening in a traumatic scene is the look of someone who knows that he or she can neither understand nor prevent a senseless act of victimization. Zizek adds that this helpless gaze "makes us all feel guilty" (211).
17. In his essay "The Cinema Animal," Geoffrey Hartman discusses this connection between film representation and murder (61–73).
18. My goal here is not to ridicule my students; rather, I want to articulate the major ways that people block themselves from interpreting historical and cultural events.
19. The Great Books theory of literature and culture still dominates many English departments and the conservative call to return to traditional texts and values.
20. My four-part rhetorical model is derived from combining Aristotle's rhetorical schema with Lacan's theory of subjectivity and A. J. Greimas's "semiotic square." For a detailed analysis of this structure, see my *Between Philosophy and Psychoanalysis* (78–83).
21. For an articulation of these different central roles in the Holocaust, see LaCapra's *History and Memory After Auschwitz* (28–32).
22. While I do agree with some of Daniel Goldhagen's arguments in his *Hitler's Willing Executioners*, his obsession with blaming every German for intentionally killing Jews blinds him from seeing many other factors that helped make the Holocaust possible. By often failing to distinguish

the leaders from the bystanders, perpetrators, and collaborators, he overlooks many of the psychological factors that connect the Nazis to our own culture. In fact, his central driving force seems to be to tell us that we have no connection to this German killing machine.

Chapter 4

1. In the next chapter I examine Freud's theory of jokes and how it relates to the critical analysis of popular culture.
2. I do not want to discount the importance of emotion and empathy in the public responses to popular culture. What I do want to argue is that these affective responses can at times block the critical analysis of culture. Consequently, they must be considered necessary but not sufficient modes of response.
3. This glorification of a personal emotional response to the pain of others can be derived from a faulty understanding of psychotherapy. Moreover, this mode of empathy has become a major form of popular entertainment.
4. For more on this rhetoric of assimilation, see Sandor Gilman's *Jewish Self-Hatred*.
5. This discussion group can be accessed on the Web at: http://post.messages.yahoo.com/bbs?action=l&tid=hv1800019119f0&sid=22198844&prop=movies:&pt=movies&p=movies.yahoo.com/shop%3fd=hv%26id=1800019119%26cf=info.
6. In many ways Freud's combination of high theory and low culture anticipates a major driving force behind postmodern culture. For more on the combination of high and low culture in postmodernity, see Frederic Jameson's *Postmodernism*.
7. This site can be found at: http://www.epinions.com/movie-review-2945-39C0EEF-38D1D893prod4?temp=comments.html.
8. Many people would argue that the film *Shoah* does successfully bridge this gap between historical knowledge and the failures of representation.
9. The Web site for the Los Angeles Museum of Tolerance (http://www.museumoftolerance.com) is an excellent source for material on the Holocaust.
10. See Mark Bracher, *The Writing Cure* (152).
11. Throughout his *Radical Pedagogy*, Mark Bracher discusses this connection between idealization and transference.
12. Along with the growing number of faculty who teach outside of the tenure system comes a corresponding increased reliance on student evaluations to judge these non-research positions. In turn, the pressure

to get good student evaluations may motivate teachers to stay away from difficult and critical subject matter.
13. For more on the negative effects of ignoring and excluding affect in education, see Lynn Worsham's "Going Postal."

Chapter 5

1. For an insightful analysis of this backlash rhetoric against the teaching of tolerance, see John K. Wilson's *The Myth of Political Correctness*. See also Dinesh D'Souza's *Illiberal Education* as a strong example of the rhetorical reversal attacking the teaching of tolerance. Finally, a prime example of the conservative effort to demonize progressive teachers can be found in David Horowitz's *The Professors*. I discuss aspects of Horowitz's work in Chapter 6 of this book.
2. Elisabeth Young-Bruehl's *The Anatomy of Prejudice* offers a broad analysis of the history of prejudice in Western culture.
3. I would argue that Woody Allen is the master of depicting the maternal superego through the assimilation of internalized anti-Semitism. Allen also uses decontextualized representations of the Holocaust in his work.
4. Throughout this work, I am using the notion of assimilation to account for the central defining process of postmodern culture. Here, assimilation not only refers to the ways that minority cultures are asked to fit into the dominant cultures but also the technological methods that allow for the reconstruction of history and narrative discourses.
5. For a discussion of the multiple meanings of circumcision in Western culture and Judaism, see Daniel Boyarin's *A Radical Jew* (67–68, 230–31, 112–13, 225–26).
6. Sandor Gilman's *Jewish Self-Hatred* provides a detailed discussion of the history of internalized anti-Semitism.
7. In the movie, one of the young characters confuses Jesus with the clitoris as the key to salvation and making a female happy.
8. While political correctness is often associated in popular culture with the supposedly oppressive power of minority groups and academic liberals, this film reveals some of the hidden anti-Semitic sources for the condemnation of these groups.
9. In researching the Web for references to *South Park*, I was shocked to discover that there were literally thousands of sites that were dedicated to distributing information about this show. For the interviews with the writers of *South Park*, see http://canniballovers.com/christy/frames/articles.html.

10. This interview with Matt Stone and Trey Parker can be found in the online version of The Jewish Student Press Service at: http://www.jsps.com/stories/southpark.shtml.
11. Throughout this work, I use Freud's book on jokes as the fundamental text defining a Jewish cultural studies approach to popular culture.
12. For an extended analysis of the Jewish aspects of Freud's jokes, see Sandor Gilman's *Jewish Self-Hatred* (261–69).
13. While Freud distinguishes among the first, second, and third parties of the joke, he does not differentiate between the object of the joke and the object's relation to the social censor.
14. I know that most teachers shy away from being critical of students. It is necessary, however, to point out the ways in which students and other people resist interpreting many aspects of popular culture.
15. Throughout Mark Bracher's *Radical Pedagogy*, we find a strong argument for helping students to acknowledge the diverse aspects of their own identities.
16. See Sartre's *Anti-Semite and Jew* (110–11).

Chapter 6

1. One of the most interesting aspects of right-wing radio talk shows is that the hosts constantly position themselves to be the victims of the liberal press and liberal politicians. The master of this reversal is Rush Limbaugh.
2. In 2005 at UCLA, a conservative alumni group offered students money to videotape their teachers to catch the faculty in the act of promoting a left-wing agenda. This action was soon discredited, but it represents just the tip of the iceberg of the new conservative effort to intimidate faculty. For an example of the conservative attack on faculty, see David Horowitz's Web site at: www.frontpagemag.com.
3. By attacking public institutions of higher education for being hotbeds of Leftist indoctrination, conservative groups can justify the defunding of these "failed" institutions. Moreover, this culture war often functions to hide the economic restructuring of American universities and colleges. Thus, the general public is often exposed to the fight between conservatives and liberals in higher education, but that same public is unaware of how these institutions have been downsized and corporatized.
4. Daniel Pipes's Middle East Forum Web site is a strong example of this conservative effort to attack academics for being anti-American, anti-Israel, and anti-Semitic. His site can be accessed at: http://www.danielpipes.org.

5. For a strong example of a conservative college student's internalization of neoconservative rhetoric, see Ben Shapiro's *Brainwashed*.
6. Throughout his *Radical Pedagogy*, Mark Bracher argues for this type of non-polarizing discourse.
7. See Gerald Graff's *Clueless in Academe*.
8. See Mark Bracher's *Radical Pedagogy*.
9. Another text my students read to analyze the rhetoric of extreme political positions on this topic is Norman Finkelstein's *The Holocaust Industry*, which represents an extreme example of internalized anti-Semitism coupled with the twin goals of questioning the American support for Israel and attacking people who fight for Holocaust compensation. In fact, Finkelstein begins his book by arguing that the American interest in the Holocaust serves to place a powerful Israel and successful American Jews in the position of being victims (3). Therefore, in this left-wing critique of American and Israeli politics, the conservative rhetorical reversal of victimhood is assimilated in order to posit that by claiming to be victims, Jews and Israelis can remove themselves from any type of criticism, while all their actions become justified (3). Finkelstein not only thinks that the "exploitation and falsification" of the Nazi genocide is used to justify the "criminal policies of Israel" but also insists that the main goal of the current Holocaust industry is to "extort money from Europe" (7–8). This incredibly inflammatory rhetoric is employed throughout his book, and it points to a strong instance of internalized anti-Semitism, for his main argument appears to be that money-hungry Jewish American lawyers are turning the Holocaust into an "extortion racket." This claim thus repeats the classic anti-Semitic stereotype that all Jews are legalistic materialists who just want to exploit other people and cash in on everyone's sympathy.

Students need to analyze the rhetoric in books like Finkelstein's so that they can examine the rhetoric of the Left and the Right in a nonpolarizing way. In fact, it is evident that the extreme rhetoric on both sides of the political spectrum tends to mimic the same set of universalized claims, and thus it is important to help students see that a shared system of false generalizations exists in American politics, media, and education. For example, underlying Finkelstein's central argument is the universalized claim that since we now live in a society where Jewish people are treated with equality and tolerance, anti-Semitism can only be the product of the Jews themselves (32). What I want my students to see in this rhetorical movement is the way that Finkelstein has to first turn to the rhetoric of universalization to deny the existence of anti-Semitism, so that he can then blame the appearance of this internalized prejudice on Jewish American organizations (33–34). By highlighting this universalizing rhetoric, I can help my students see how people on the Left and the Right now tend to employ the modern notion of universal equality and a globalized sense of a lack of identity in order to deny the real histories of prejudice affecting different ethnic groups.

Yet, what makes Finkelstein's argument liberal and not conservative is that he desires to show how Jews want to claim their victim status in order to fight affirmative action and other progressive policies favoring different minority groups (36). Here Finkelstein internalizes the anti-Semitic stereotype that Jews are so hung up on their own history of suffering that they cannot consider the suffering of others (8). Moreover, his turn to a universal logic pushes him to agree with his mother's credo that "We are all Holocaust victims" (8). Here we see that while conservatives often use the universal category to deny the differences and sufferings of minority groups, liberals often invoke universalistic claims to posit victim status for all people. However, what connects both of these logics is the idea that if everyone is a victim, then no one is.

Finkelstein's obsession with this Jewish victimhood is presented in many different ways. One of his favorite targets is Elie Wiesel, who is attacked for taking money to talk to people about the unspeakable nature of the Holocaust (45). By labeling Wiesel a "survivor-priest," Finkelstein claims that Wiesel's testimonies represent a "sacralization of the Holocaust" (45). Finkelstein goes one step further and questions the validity of Wiesel's entire discourse. He does this by listing a few inconsistencies that Wiesel has made in various places, and then he universalizes these distortions to condemn Wiesel's entire project and to question the value of every survivor's testimony (82). Finkelstein not only posits that we cannot trust these narratives but also adds that anyone who questions them will be labeled anti-Semitic and a Holocaust denier (82). Perhaps he makes this argument because he is unconsciously aware of his own anti-Semitic discourse.

Finkelstein's internalized anti-Semitism is evident in the ways that he denies the value of the Holocaust survivors. In relation to Wiesel, Finkelstein accuses this survivor of trying to divide the world between the ever-innocent, suffering Jew and the ever-guilty Gentile (54). This opposition is then employed to show how Wiesel positions himself as embodying the Holocaust in order to receive unconditional support for Israel and other Jewish interests (55). Once again, Finkelstein represents the Jewish survivor as partaking in a giant conspiracy to make Gentiles feel guilty and to gain support for secret Jewish moneymaking schemes.

There is a certain truth to the idea that various groups and politicians use the threat of anti-Semitism and Holocaust denial to silence critics of Israel and the United States. Nevertheless, what I hope my students learn from analyzing Finkelstein's rhetoric is the way that real issues are attached to universalized claims that render the initial premises suspect. A strong example of this rhetorical move can be found in

Finkelstein's argument that much of the literature documenting Hitler's Final Solution is "worthless scholarship" and that the field of Holocaust studies is "replete with nonsense" (55). In order to make the extreme universalizing claim that Holocaust scholarship is basically "sheer fraud," he turns to a couple of examples of faked Holocaust testimonies, and then he applies a false universal logic to question all Holocaust literature (55–58). Moreover, he concludes that because many Holocaust scholars did not immediately detect the fact that these books were frauds, the Holocaust industry is built on "a fraudulent misappropriation of history" (61).

While Finkelstein is extremely hard on any real Holocaust survivor's ability to tell the whole truth about this traumatic period of history, he quickly dismisses the Holocaust deniers as having no more effect on the American public than the flat earth society (68). Moreover, he adds that given the "nonsense" churned out by the Holocaust industry, it is a wonder that more people do not doubt or deny the Holocaust (70). The only people that Finkelstein seems to pity are the Nazis, who he feels are often portrayed in Holocaust testimonies as being violent and sadistic (58).

This strange defense of the Nazis leads up to Finkelstein's main concern, the Jewish plot to "extort" money from Europe. In this section of his book, we see why he must deny the validity of the Holocaust survivors' testimonies. For, what he wants to prove is that the claims for reparations have been inflated and that they depend on a false calculation regarding the actual number of living survivors (83). The reason why Finkelstein seems to be so interested in this question of reparations is that his mother was a survivor, and he is bothered by the fact that she received merely $3,500 from the German government (85). His anger is in part derived from the "fact" that many people who were not even real victims of the camps received lifetime pensions from Germany worth hundreds of thousands of dollars (85).

While we can sympathize with his mother's plight, we have to examine the role this plays in his entire project, for the source of his own assimilated anti-Semitism appears to be in part connected to the failure of his mother to get what he thinks was owed to her. Out of sheer frustration, he turns his anger away from the German authorities and blames Jewish people for his mother's plight. He then is pushed to participate in his own brand of Holocaust denial by questioning the value and authenticity of the survivors' testimonies. But this denial is not enough, and he must question the very number and validity of the survivors themselves. In fact, since his mother did not get her fair share, Finkelstein appears bent on demonizing the whole process of war reparations. He claims that the "Holocaust restitution racket" has been one of the greatest robberies of human history (94). He sees the spread of

this conspiracy, or what he calls the "shakedown," as an ever-expanding project whose next target is the poor countries of Eastern Europe (130). Furthermore, he blames the current rise of anti-Semitism in these countries on the reckless Holocaust extortion industry (130).
10. It is important to stress that Sartre's *Anti-Semite and Jew* also denies Jewish people multiple avenues for ethnic and religious identity, and the result of this denial is his argument that anti-Semitism is the only real source for Jewish identity.
11. The success of many Jewish Americans in the media and other public places has often worked to feed anti-Semitic myths and hide the presence of past and present unsuccessful Jewish Americans.
12. In one of my classes, students analyzed the rhetoric of radio talk shows just after Joe Lieberman was chosen to be Al Gore's vice presidential running mate in 2000. Here are some of the comments my students recorded and examined:

1) Won't he [Lieberman] be controlled by Israel?
2) Of course people like him; they control the press.
3) Why is it that when a Christian is religious, he's branded an extremist by the media, but when a Jew is religious, he's a great man?
4) The whole thing is a liberal plot. Clinton convinced Al to play the "Jew card," so people in New York won't attack Hillary for being anti-Semitic.
5) The liberals like to pick victims, so they can say that the liberal government has the cure.
6) What is this love affair that liberals have with minorities?
7) I like the guy, but I think that other people will vote against him because they are less tolerant than I am.
8) It might be good for the Jews, but is it good for the country?
9) Gore picked him because if people attack Gore, he can call them anti-Semitic.
10) Now Gore has a great excuse if he loses.

In looking back over these comments, the first thing that becomes glaringly obvious is how the political discourse has changed after 9/11. Likewise, none of these callers on radio talk shows came right out and said that they did not trust Jewish people. However, many of these callers repeated the most basic anti-Semitic myths: Jews secretly control the world through their control of the media; Jews see themselves as eternal victims who everyone is supposed to feel sorry for; Jews are always plotting and lying; and Jews blame their problems on everyone else.

In these statements, "the Jew" is positioned as being both a "victim" *and* a powerful controller of culture and politics. Like the Nazis who saw the Jews as subhuman animals *and* people who secretly controlled

German politics and banks, Jews in America are often represented as being a disempowered minority *and* a powerful member of the majority at the same time. Furthermore, some people claim that there is no anti-Semitism in America, yet they also claim that the Jews get away with things because they can label any attack on them as being anti-Semitic. Of course, this contradiction between affirming and denying the current presence of anti-Semitism is overcome by the idea that anti-Semitism exists only in the minds of Jewish people.

Not only do many of these statements hold to the idea that Jews invent anti-Semitism but also many of these prejudices cling to the idea that liberals need minorities in order to have a reason to expand the government and increase their own power. According to this logic, Jews position themselves as minority victims so that they can then take power by helping their fellow victims. This representation of American Jews as powerful victims is often attached to the question of the American support for the state of Israel. Instead of just coming right out and saying that people are afraid of the political and economic clout of American Jews, many radio talk show callers and hosts focus on Israel and its tie to American Jews.

In the case of Joe Lieberman, several callers wondered whether he could be open-minded because he would naturally have to support his own people and the state of Israel. Of course, this argument assumes that all Jewish people are uncritical of Israel and cling to a separate political agenda. In fact, many callers implied that since the Jews control the media, Lieberman would not be criticized. This theory is in turn extended to the claim that Gore picked Lieberman precisely to protect himself from any future criticism on any subject. For, many callers posited that you cannot criticize a Jew or a friend of a Jew without being labeled anti-Semitic. One caller expressed this opinion through the following tortured logic: "Jews have always made Americans feel ashamed for being anti-Semitic even though Americans are the most tolerant people around. Gore decided to pick Lieberman because he knew how ashamed Americans have been made to feel about Jews." This opinion is based on the faulty claim that Jews invented anti-Semitism. This denial of American anti-Semitism is then attached to the idealization of American tolerance. The result of these two claims is that the victims of anti-Semitism are not only blamed for its existence but also attacked for using it to their advantage.

One of the arguments that many of these veiled anti-Semitic claims cling to is the idea that the real victims in our culture are white male Christians. According to this logic, the Jews control the media, politics, and the economy, and their main goal is to retain power by making everyone else feel guilty and by absolving themselves of any type of guilt or criticism. This idea also implies that the Jews see themselves as all good and the Christians as all bad. We see this logic applied in the argument that Gore needed to purify himself by attaching himself to a religious Jew. In fact, it is not even Gore who had to be purified but former president Bill Clinton, whose stain had to be removed through some act of Jewish purification. In this strange situation, Lieberman the Jew becomes the savior for Gore the Christian.

13. In *Writing Prejudices*, I provide a detailed analysis of how sexism, racism, homophobia, and ethnocentrism are all related and different.
14. In his book *History and Memory After Auschwitz*, Dominic La Capra does a good job analyzing the connection between Jewish identity and Jewish victim status after the Holocaust.
15. In Chapter 3 of this book I present a four-part model that relates the idealization of the leader to the identification with a victim status, coupled with the assimilation of prejudices and the universalization of indifference.
16. For an extended analysis of this false universalizing logic of tolerance, see Stanley Fish's *There's No Such Thing as Free Speech*.
17. A proof of the rhetorical nature of political idealizations can be found in the fact that before 9/11, many conservatives attacked the power of liberal Jews to control American foreign policy, while many liberals defend the American investment in Israel. Now, however, we find that many conservatives are idealizing Israel, and many liberals are demonizing Israel.
18. I would argue that Israel has suffered from being positioned in the American press as both a Fascist neo-Nazi military state and as a country full of hopeless victims of the Holocaust.
19. Some of Novick's animosity toward the victimhood status of the Jews can be derived from a conservative belief in self-help and self-determination.
20. In his *Selling the Holocaust*, Tim Cole falls into the same trap of calling the Holocaust a myth and stressing the fictional representation of history over the actual experience of history (1–19).
21. This equation between ethnic difference and victim status has been an effective conservative and moderate argument employed to help defund a diverse array of social policies.
22. By blaming the victims of poverty for poverty and the recipients of welfare for the welfare state, politicians are able to hide all structural and economic differences behind the question of self-improvement.
23. I am arguing here that just as the popular culture often seeks to hide social messages behind a veil of non-meaning, academics participate in this same process through their use of negative arguments.

Works Cited

Adorno, Theodor W., and Max Horkheimer. *Dialectic of Enlightenment.* Translated by John Cumming. New York: Continuum, 1972.
Alcorn, Marshall. *Changing the Subject in English Class.* Carbondale: Southern Illinois University Press, 2002.
Aristotle. *On Rhetoric.* Translated by George Kennedy. New York: Oxford University Press, 1991.
Bartov, Omer. "Spielberg's Oskar: Hollywood Tries Evil." In *Spielberg's Holocaust: Critical Perspectives on Schindler's List*, by Yosefa Loshitzky, editor, 41–60. Bloomington: Indiana University Press, 1997.
Baudrillard, Jean. *The Vital Illusion.* New York: Columbia University Press, 2001.
Beckwith, Barbara. "The U.S. Holocaust Museum: Why Christians Should Go." The History Place. Accessed online March 9, 2001. http://www.historyplace.com/ pointsofview/beckwith.html.
Berman, Jeffrey. *Risky Writing: Self-Disclosure and Self-Transformation in the Classroom.* Amherst: University of Massachusetts Press, 2001.
Biale, David, Michael Galchinsky, and Susannah Heschel. *Insider/Outsider: American Jews and Multiculturalism.* Berkeley: University of California Press, 1998.
Bok, Derek. *Our Underachieving Colleges.* Princeton: Princeton University Press, 2006.
Boyarin, Daniel. *A Radical Jew: Paul and the Politics of Identity.* Berkeley: Universityof California Press, 1994.
Bracher, Mark. *Radical Pedagogy.* New York: Palgrave, 2006.
———. *The Writing Cure.* Carbondale: Southern Illinois University Press, 1999.
Caruth, Cathy. *Unclaimed Experience: Trauma, Narrative, and History.* Baltimore: Johns Hopkins University Press, 1996.
Cole, Tim. *Selling the Holocaust: From Auschwitz to Schindler; How History Is Bought, Packaged, and Sold.* New York: Routledge, 1999.
Descartes, René. *Discourse on Method.* New York: Penguin Books, 1987.
D'Souza, Dinesh. *Illiberal Education: The Politics of Race and Sex on Campus.* New York: Free Press, 1991.

Works Cited

Ellsworth, Elizabeth. "The U.S. Holocaust Museum as a Scene of Pedagogical Address." *Symploke* 10, no. 1–2 (2002): 13–31.

Felman, Shoshana, and Dori Laub. *Testimony: Crises of Witnessing in Literature, Psychoanalysis, and History*. New York: Routledge, 1992.

Finkelstein, Norman G. *The Holocaust Industry: Reflections on the Exploitation of Jewish Suffering*. London: Verso, 2000.

Fish, Stanley. *There's No Such Thing as Free Speech*. New York: Oxford University Press, 1994.

Freud, Sigmund. *Beyond the Pleasure Principle*. Translated by James Strachey. New York: W. W. Norton, 1961.

———. *Civilization and Its Discontents*. Translated by James Strachey. New York: W. W. Norton, 1961.

———."Creative Writers and Day-Dreaming."In *The Freud Reader*, by Peter Gay, editor, 426–43. New York: Norton, 1989.

———. *The Ego and the Id*. Translated by James Strachey. New York: W. W. Norton, 1961.

———. *Future of an Illusion*. Translated by James Strachey. New York: W. W. Norton, 1961.

———. *Group Psychology and the Analysis of the Ego*. Translated by James Strachey. New York: W. W. Norton, 1961.

———. *Jokes and Their Relation to the Unconscious*. New York: W.W. Norton, 1960.

———. "Mourning and Melancholia." In *Complete Psychoanalytical Works*, by James Strachey, editor (24 vols.), 14:237–58. London: Hogarth Press and the Institute of Psycho-Analysis, 1955.

Friedman, Lester D., and Brent Notbohm. *Steven Spielberg: Interviews*. University Press of Mississippi, 2000.

Fromm, Erich. *Escape from Freedom*. New York: Avon Books, 1965.

Gergen, Kenneth. *The Saturated Self: Dilemmas of Identity in Contemporary Life*. New York: Basic Books, 1991.

Giami, Alain. "Counter-Transference in Social Research: Beyond George Devereux." *Papers in Social Research Methods—Qualitative Series*, no. 7 (2001). Accessed online June 4, 2002. http://www.ethnopsychiatrie.net/giami.htm.

Gilman, Sandor. *Freud, Race, and Gender*. Princeton: Princeton University Press, 1993.

———. *Jewish Self-Hatred: Anti-Semitism and the Hidden Language of the Jews*. Baltimore: Johns Hopkins University Press, 1986.

Goldhagen, Daniel Jonah. *Hitler's Willing Executioners: Ordinary Germans and the Holocaust*. New York: Knopf, 1996.

Gourevitch, Philip. "The Memory Thief." *The New Yorker* (June 14, 1999): 48–68.

———. "What They Saw at the Holocaust Museum." *New York Times Magazine* (February 12, 1995): 44–45.

Graff, Gerald. *Clueless in Academe*. Chicago: University of Chicago Press, 1994.
Hansen, Miriam Bratu. "*Schindler's List* Is Not *Shoah*: Second Commandment, Popular Modernism, and Public Memory." In *Spielberg's Holocaust: Critical Perspectives on Schindler's List*, by Yosefa Loshitzky, editor. Bloomington: Indiana University Press, 1997.
Hartman, Geoffrey H. "The Cinema Animal." In *Spielberg's Holocaust: Critical Perspectives on Schindler's List*, by Yosefa Loshitsky, editor. Bloomington: Indiana University Press, 1997.
Heidegger, Martin. *The Question Concerning Technology*. Translated by William Lovitt. New York: Harper and Row, 1997.
Horowitz, David. *The Professors: The 101 Most Dangerous Academics in America*. Washington, DC: Regnery Publishing, Inc., 2006.
———. Web site: www.frontpagemag.com.
Horowitz, Sara. "But Is It Good for the Jews? Spielberg's Schindler and the Aesthetics of Atrocity." In *Spielberg's Holocaust: Critical Perspectives on Schindler's List*, by Yosefa Loshitzky, editor, 119–39. Bloomington: Indiana University Press, 1997.
Jameson, Frederic. *Postmodernism: Or, The Cultural Logic of Late Capitalism*. Durham: Duke University Press, 1991.
Keneally, Thomas. *Schindler's List*. New York: Simon and Schuster, 1982.
Lacan, Jacques. *The Four Fundamental Concepts of Psychoanalysis*. Translated by Alan Sheridan, Jacques Alain Miller, editor. New York: Norton, 1977.
———. *L'envers de la psychanalyse*. Paris: Seuil, 1991.
LaCapra, Dominick. *History and Memory After Auschwitz*. Ithaca: Cornell University Press, 1998.
———. *Representing the Holocaust: History, Theory, Trauma*. Ithaca: Cornell University Press, 1994.
———. *Writing History, Writing Trauma*. Baltimore: Johns Hopkins University Press, 2001.
Landsberg, Alison. "America, the Holocaust, and the Mass Culture of Memory: Toward a Radical Politics of Empathy." *New German Critique: An Interdisciplinary Journal of German Studies* 71 (Spring–Summer 1997): 63–87.
Langer, Lawrence. *Holocaust Testimonies: The Ruins of Memory*. New Haven: Yale University Press, 1991.
Linenhal, Edward. *Preserving Memory: The Struggle to Create America's Holocaust Museum*. New York: Penguin Group, 1995.
Loshitzky, Yosefa. "Holocaust Others: Spielberg's Schindler's List Versus Lanzmann's Shoah." In *Spielberg's Holocaust: Critical Perspectives on Schindler's List*, by Yosefa Loshitzky, editor, 104–18. Bloomington: Indiana University Press, 1997.

McBride, Joseph. *Steven Spielberg: A Biography*. New York: Simon and Schuster, 1997.
Novick, Peter. *The Holocaust in American Life*. New York: Houghton Mifflin, 1999.
Palmer, Parker. *The Courage to Teach*. San Francisco: Jossey-Bass, 1998.
Pascal, Michael. "Steven Spielberg: Why I Made Schindler's List." *Queen's Quarterly* 101, no. 1 (Spring 1994): 27–33.
Patraka, Vivian. "Situating Genocide and Difference: The Cultural Performance of the Term Holocaust in U.S. Public Discourse." In *Jews and Other Differences: The New Jewish Cultural Studies*, by Jonathan Boyarin and Daniel Boyarin, editors, 54–78. Minneapolis: University of Minnesota Press, 1997.
Perry, George. *Steven Spielberg*. New York: Thunder Mouth, 1998.
Samuels, Robert. *Between Philosophy and Psychoanalysis*. New York: Routledge, 1993.
———. *Writing Prejudices*. New York: SUNY Press, 2000.
Santer, Eric. *Stranded Objects. Mourning, Memory, and Film in Postwar Germany*. Ithaca: Cornell University Press, 1990.
Sartre, Jean-Paul. *Anti-Semite and Jew*. New York: Shocken Books, 1948.
Shapiro, Ben. *Brainwashed*. World Net Daily Books, 2004.
Tingle, Nick. *Self-Development and College Writing*. Carbondale: Southern Illinois University Press, 2004.
Walkerdine, Valerie. *Daddy's Girl: Young Girls and Popular Culture*. Cambridge: Harvard University Press, 1997.
Wilson, John K. *The Myth of Political Correctness*. Durham: Duke University Press, 1995.
Worsham, Lynn. "Going Postal: Pedagogic Violence and the Schooling of Emotion." *JAC: A Journal of Composition Theory*18 (1998): 213–45.
Wyatt, Jean. *Risking Difference*. Albany: State University of New York Press, 2004.
Yahoo! Movies: Life Is Beautiful. Accessed June 1, 2000 http://post.messages.yahoo.com/bbs?action=l&tid=hv1800019119f0&sid=22198844=movies:&pt=movies&p=movies.yahoo.com/shop%3fd=hv%26id=1800019119%26cf=info.
Young-Bruehl, Elisabeth. *The Anatomy of Prejudices*. Cambridge: Harvard University Press, 1996.
Zelizer, Barbie. *Remembering to Forget*. Chicago: University of Chicago Press, 1995.
Zizek, Slavoj. *Looking Awry: An Introduction to Jacques Lacan through Culture*. Cambridge: MIT Press, 1992.
———. *The Metastases of Enjoyment*. New York: Verso, 2006.
———. *The Plague of Fantasies*. New York: Verso, 1997.
———. *The Sublime Object of Ideology*. New York: Verso, 1989.

INDEX

academic freedom, 129, 132, 151
Adorno, Theodor W., 17
Alcorn, Marshall, 15, 156n12
alienation, 17, 69
American Jews, 11, 52–53, 135, 139–42, 163n9, 167n12
anti-Semitism, 1, 11, 30–31, 42, 50–53, 79, 83, 85, 99, 113, 115, 119, 124–25, 128–29, 132, 134–43, 149–50, 161n3, 161n6, 163n9, 166n10, 167n12
Aristotle, 4, 8, 49, 155n7
assimilation, ix, xi, 3–4, 16, 19–24, 25–26, 29, 36, 39, 42, 43, 46, 51, 52–55, 56, 59, 65–69, 74–75, 78, 79–85, 87, 95–96, 104, 108, 109, 111, 113, 115, 118–19, 121, 125–29, 140, 142, 149, 150, 157n7, 160n4, 161n3, 161n4, 168n15
authoritarianism, 4
automodernity, 78

backlash rhetoric, x, 2, 5, 8, 25, 78, 100, 112, 137, 150, 155n1, 161n1
Bartov, Omer, 65, 157n4
Baudrillard, Jean, 159n15
Benigni, Roberto, 25, 98, 99, 100–102
Berman, Jeffrey, 6, 155n8, 156n10
Bok, Derek, 7, 155n9
Borsh-Jacobson, Mikkel, 9
Boyarin, Daniel, 161n5

Bracher, Mark, xi, 6, 14, 34, 160n10, 160n11, 162n15, 163n6
Butler, Judith, 9

Caruth, Cathy, 158n14
Cole, Tim, 155n4, 156n3, 168n20
conscience, 9, 13–14, 21
critical self-reflection, ix, 6, 8, 26, 35, 47, 69, 85–86
critical thinking, xi, 3, 6, 10, 19, 53, 56, 59–61, 85–86, 87, 91, 96–97, 112–13, 132, 133–34, 148
cultural relativism, 10, 83, 133
cultural studies, 27, 48, 49, 53, 96, 119, 157n1, 162n11

defense mechanisms, x, 3, 5–24, 25, 29, 35, 36–38, 42, 47, 49, 53, 54, 56, 59, 64, 66, 71, 73, 74–78, 80–81, 84–85, 92–93, 95–96, 104, 109, 111, 133, 138, 139, 147, 151, 155n5, 156n10, 158n8
de-idealization, 6, 14, 15, 50, 105
democracy, 30, 51, 56, 89, 111
Descartes, René, 17, 156n18
D'Souza, Dinesh, 155n1, 161n1
dream analysis, 23–24, 48, 60, 96

educational settings, 2–3, 5–7, 14–15, 18–19, 22–23, 24, 27, 30–31, 34, 36, 38, 39, 40, 43, 45, 46, 54–56, 60, 81, 85, 87,

92, 95, 104, 105–6, 109, 117, 129, 133, 151, 152
Ellsworth, Elizabeth, 156n2
empathy, 9–13, 14, 19, 32–35, 39, 44, 45, 67–68, 73–74, 75, 77, 78, 80, 91–92, 94, 96–97, 139, 160n2, 160n3
enlightenment, 16–17, 142
ethics, xi, 3, 7, 14, 15, 17, 29–30, 44, 45, 48, 62, 68, 73, 80
ethnicity, ix, 4, 17, 21, 43, 50–53, 66, 74, 77, 82, 101, 111, 113, 118–22, 124–26, 134–35, 137, 140, 142, 164n10, 168n21
ethnography, 31–32, 77

fascism, 4, 94, 125
Felman, Shoshana, 12–13
fetish, 21
Finkelstein, Norman, 155n4, 156n3, 163n9
Fish, Stanley, 120, 129, 168n16
free association, x, 7, 23, 32, 63–64, 152
free speech, 103–4, 115–16, 119–21, 127, 129, 151, 168n16
Freud, Sigmund, 15, 19, 23, 35, 48, 49, 60, 96, 125, 129, 160n1, 160n6
  Beyond the Pleasure Principle, 16, 18, 20–21
  "Creative Writers and Day-Dreaming," 60, 157n2
  The Ego and the Id, 9, 21–22
  Future of an Illusion, 15
  Group Psychology and the Analysis of the Ego, 9, 10–11, 13–14, 79
  Jokes and Their Relation to the Unconscious, 20, 25, 88, 117–19, 121–22, 125
  "Mourning and Melancholia," 11

Friedman, Lester D., 158n7
Fromm, Erich, 13
fundamentalists, 4, 8, 149

generalization, 12, 16, 95, 106, 113, 122, 126, 163n9
Giami, Alain, 107
Gilman, Sandor, 20, 124–25, 160n4, 161n6, 162n12
Global War on Terror, x, 5, 26, 148, 149–50
globalization, ix, 4, 7, 19, 26, 27, 52, 73, 83, 85, 89–91, 111, 112, 116, 143, 148, 149–50
Goldhagen, Daniel Jonah, 79, 159n22
Gourevitch, Philip, 36–40, 157n6
grades, 55–56, 64, 72, 152
Graff, Gerald, 163n7
Great Books, 76, 78, 159n19
guilt, 14, 88, 132, 139, 140, 142, 144, 159n16

Hansen, Miriam Bratu, 65, 157n6, 158n10
Hartman, Geoffrey H., 157n6, 159n17
Holocaust, ix–xi, 1–4, 10–13, 15, 17, 19, 22, 26, 29–39, 42–54, 56–57, 59, 61–62, 67–73, 78–80, 83–84, 87–88, 91, 94, 96–104, 107–8, 111, 114, 120, 123–24, 128–29, 132, 134–44, 149–52, 155nn4–5, 156nn2–3, 157nn5–6, 158n12, 159n16, 159nn21–22, 160n9, 161n3, 163n9, 168n14, 168n18, 168n20
Horkheimer, Max, 17
Horowitz, David, x, 26, 143–46, 150–51, 155n3, 161n1, 162n2
Horowitz, Sara, 66, 158n12
hysteria, 10

INDEX

idealization, ix, xi, 3–11, 13–16, 24, 25, 26–27, 29, 43–44, 46–47, 49–54, 59, 61–64, 69–76, 78–83, 85, 87, 92–94, 96, 98, 103–6, 108, 111, 115–16, 129, 132–34, 139, 148, 149, 151–52, 156n15, 160n11, 167n12, 168n15
identification, ix, x, 3–4, 8–16, 19, 21, 24–27, 29, 32–35, 38–40, 43–47, 51, 53, 56, 59, 63, 67–69, 73–75, 77–83, 87, 91–94, 96, 97, 100, 104, 119, 125, 134–40, 149–51, 156n11, 156n13, 168n15
identity fixation, 3, 10–12, 17, 22, 51, 53, 75, 80, 82–83, 92, 113, 124, 131, 135, 140, 142, 149, 151
identity politics, ix, 4, 10–12, 17, 25, 26, 51, 82, 83, 119, 126, 132–37, 140
imaginary, 93
indifference, 7, 17, 18–19, 30, 33, 44, 56, 75, 80, 83, 149, 168n15
internalized racism, 22, 26, 52, 83–84, 111, 113, 117, 119, 121, 125, 138, 142, 149–51
intolerance, 1–2, 8, 38, 43, 50, 73, 84, 111–14, 118, 126, 127–28, 131, 139, 152, 155n2
Israel, x, 132, 134, 137–39, 143–46, 149–51, 162n4, 163n9, 167n12, 168n17

Jameson, Frederic, 160n6
Jewish identity, 11, 53, 113, 117, 124–25, 135, 137, 140, 166n10, 168n14
jokes, theory of, 20–21, 25, 88, 117–21, 125, 127, 128, 160n1

Kristeva, Julia, 9

Lacan, Jacques, 9, 16, 61–62, 64, 93, 104, 156n17, 158n9, 159n20
LaCapra, Dominick, 11–13, 155n6, 156n13
Lanzman, Claude, 12
liberal, 9, 26, 51, 83, 112, 116, 129, 132, 150, 161n8, 161n8, 162n1, 162n3, 164, 166n12, 168n17
Life is Beautiful, 25, 87–89, 95–102
Linenthal, Edward, 30, 42–46
Loshitzky, Yosefa, 66, 157n5
lost object, 9, 11, 15, 16, 18

mass media, 41, 52, 126
McBride, Joseph, 157n3
media literacy, 27, 50
melancholia, 9, 11, 15
modernity, 15, 16–18, 25, 46, 56, 78–79, 82, 84, 89–90, 106–7, 125–26, 156n1, 163n9
mourning, 11, 15, 18, 156n12
multiculturalism, 4, 38, 77–78
Museum of Tolerance, 24, 29, 36, 46–53, 160n9

narcissism, 13, 27, 69, 94
national socialism, 128
nationalism, 26, 149
negation, 21, 35
new media, 78
9/11, x, 4, 26, 143–45, 149
Novick, Peter, 26, 53, 134–43, 155n4, 156n3, 168n19

objectivism, 18
objectivity, 9–10, 18, 25, 83, 106, 108–9
online writing, 8, 32, 97, 105, 119, 143

Palmer, Parker, 18
Pascal, Michael, 157n3
pedagogy, ix–xi, 3, 5–7, 10, 12, 14, 15, 23, 25–26, 30, 33–35, 42–45, 47, 50, 53–56, 59–62, 64, 66–67, 71, 74, 80, 84–85, 87, 92–93, 95–96, 104, 106, 119, 122, 132–33, 138, 140, 144–45, 151–52, 153, 157n1
personal writing, 6, 23–24, 54–56, 106, 108, 119
political correctness, 8, 112, 116, 119–20, 128, 155n1, 161n1, 161n8
politically incorrect, 111, 117
popular culture, ix–xi, 2, 3, 4, 7, 15, 18–27, 29–31, 34, 43, 46, 48, 50, 51, 52, 59–61, 66–67, 70–75, 78, 80–85, 87–90, 93–97, 101–4, 108, 111–28, 139–42, 149–53, 160nn1–2, 161n8, 162n11, 162n14, 168n23
postmodernity, 24, 25, 28, 42–43, 56, 66–67, 69, 77–78, 84, 94, 97, 113, 115, 126, 128, 233, 140, 150, 156n1, 158n14, 160n6, 161n4
prejudice, 1–2, 3–5, 7–8, 11, 20–22, 24, 24–27, 30, 35–39, 42, 46–53, 65–66, 79, 82–84, 89–90, 95, 108, 111–14, 116–18, 121–22, 126–28, 131, 133–35, 138–40, 149–50, 157n4, 161n2, 163n9, 168n13, 168n15
progressive educators, 2–6, 9, 14, 15, 21, 23
progressive pedagogy, 25–26, 29–30, 31, 112, 132
projection, 9, 14, 108

religion, 15, 17, 38, 61, 76, 80, 128, 140, 166n10, 166n12
repetition, 1, 16–19, 36, 39, 75, 82, 90, 141
repression, 18, 20, 23, 31–32, 35, 46, 83, 92, 106–7, 109, 117, 119, 148, 159n16
resistance, ix–xi, 3, 5–8, 10–11, 14–16, 19–20, 23–25, 29, 38–39, 47, 54, 56, 59, 61, 63, 68, 70–75, 78, 84–85, 88, 91–98, 100–1, 104–9, 112–15, 118, 129, 133, 136, 147, 148, 155n6, 158n13, 159n16
reversed racism, 8, 112
rhetoric, x–xi, 2–4, 9–10, 15–27, 29, 33, 35–36, 38–39, 42–45, 47–50, 53–54, 59, 61, 63, 64, 67–69, 71–85, 87–97, 101, 104–9, 111–18, 120, 122, 128–29, 131–48, 149–51, 155, 159, 160, 161, 163n5, 163n9, 166n12, 168n17

Santer, Eric, 156n15
Sartre, Jean-Paul, 53, 126, 139, 162n16, 166n10
Schindler's List, 25, 34, 59–85, 98, 157n3, 157n6, 158n10
science, 1, 15, 17–18, 20, 21, 25, 32, 77, 79, 82, 89–90, 106–8, 109, 125–26, 156n18, 157n4
self-destruction, 9, 66, 112
self-division, 21, 22, 35
self-hatred, 20, 52, 113, 124–26, 160n4, 161n6, 162n12
selfhood, 10, 55, 106
Shapiro, Ben, 163n5
Shoah, 12, 34, 67, 158n10, 160n8
social change, 2–3, 10, 11, 24, 26, 29, 31, 37, 44, 47, 53–54, 56,

59, 60, 73, 85, 91–92, 112, 124, 132, 148, 152
South Park: Bigger, Longer, and Uncut, 25, 111–29, 161n9
Spielberg, Steven, 25, 59, 61–75, 84–85, 157nn3–6, 158n7, 158n10, 158n13
stereotypes, 4, 5, 7, 20–22, 25, 36, 37, 39, 49, 52, 65–67, 74–75, 77, 82, 84, 95, 111–16, 121–22, 128, 139, 142, 158n10, 163n9
students, ix–xi, 2–8, 11–15, 18–20, 22–25, 29, 32, 35–38, 50, 54–56, 59–86, 87–88, 90–101, 104–9, 111–21, 124–27, 129, 132–37, 140–51, 156n10, 156n3, 157n1, 159n16, 159n18, 162nn14–15, 162n2, 163n9, 166n12
subjectivity, 6, 9–10, 18, 22, 66, 69, 74, 79, 85, 93, 106–9, 148, 152, 158n14, 159n20
symbolic, 12, 16–19, 23–24, 39, 45–46, 48, 69, 75, 77, 79, 81–82, 90, 129, 142, 150, 157n1

tax breaks, 112, 131, 150
teachers, 3–6, 12, 14, 19, 23–26, 38, 41, 44, 54–55, 59, 61, 64, 72, 87, 93–95, 105–6, 109, 114, 116, 121, 131–35, 141, 143, 145–47, 152, 155n3, 160n12, 161n1, 162n14
Tingle, Nick, 156n12
transference, 12, 25, 86, 87, 92, 104, 106–8, 155n6, 160n11
trauma, ix, 3, 4, 9, 12–13, 19, 35, 70, 112, 129, 159n16
    historical, ix, 1, 10–13, 15, 24, 26–27, 29–30, 33, 35, 37, 39, 42–48, 53, 54, 56–57, 59, 62–63, 67, 73, 82, 87–88, 91, 104, 124, 134–35, 138, 140, 149, 149–50, 152–53, 155n5, 157n1
    individual, 8, 16, 18
    as limit to representation, 11–13, 156n13, 158n14

unconscious, x, 2, 3, 5–6, 9, 14, 20, 22–24, 25, 26, 29, 40, 46, 49, 60–63, 65, 71, 77, 82–83, 84, 90–92, 101, 106–7, 109, 111, 117, 122, 137, 164
United States, x, 4, 51, 55, 123, 131–35, 138–39, 143, 146, 149–50, 164
United States Holocaust Memorial Museum, 24, 29–54, 62, 78, 141, 156n2
universalization, x, xi, 3–5, 7, 9, 12, 16–19, 24, 25, 26, 29, 31, 35–36, 38–40, 43, 45–46, 49, 51–53, 56, 59, 68–75, 77–80, 82–84, 87–91, 94, 96, 98, 104, 112, 114, 116, 125–26, 129, 132–33, 135–40, 143–44, 147–48, 150–52, 156n18, 163n9, 168n15

victim identity, x, 2–5, 8, 10, 11–13, 16, 25–27, 33, 38–39, 40, 43–45, 52, 68, 69, 75, 78, 80, 82–83, 97, 112–13, 115, 119, 124, 131–32, 134–44, 149–51, 157n8, 158n12, 159n16, 163n9, 166n12

welfare state, 9, 112, 131, 150, 168n22

Western culture, 1–2, 11, 15, 56,
    63, 126, 161n5
Wilson, John K., 155n1, 161n1
World War I, 13
World War II, 79, 114, 141
World Wide Web, ix, 3, 8, 25, 29,
    31–32, 84, 87, 95–104,
    126–28, 132, 135, 143–44,
    147, 157n6, 160n5, 160n9,
    161n9, 162n2, 162n4

Worsham, Lynn, 157n1, 161n13
Wyatt, Jean, 9–11

Young-Bruehl, Elisabeth, 157n4,
    161n2

Zelizer, Barbie, 157n8
Zizek, Slavoj, 17, 21, 89–90,
    156n13, 159n16